Current Topics in
Microbiology
124 and Immunology

Genetic Control
of the Susceptibility
to Bacterial Infection

Edited by David E. Briles

With 19 Figures

Springer-Verlag
Berlin Heidelberg NewYork Tokyo

Professor Dr. DAVID E. BRILES, Ph.D.
Cellular Immunobiology Unit of the Tumor Institute
Department of Microbiology and Pediatrics, and
Comprehensive Cancer Center
University of Alabama at Birmingham
Birmingham, Alabama 35294, USA

ISBN 3-540-16238-0 Springer-Verlag Berlin Heidelberg New York Tokyo
ISBN 0-387-16238-0 Springer-Verlag New York Heidelberg Berlin Tokyo

© by Springer-Verlag Berlin Heidelberg 1986
Library of Congress Catalog Card Number 15-12910
Printed in Germany.

Typesetting, printing and bookbinding:
Universitätsdruckerei H. Stürtz AG, Würzburg
2123/3130-543210

Preface

This series of reviews focuses on recent developments in understanding bacterial pathogenesis that have been gained by studying the genetic control of the susceptibility to particular diseases. The topics of the reviews include a description of bacterial genes that effect virulence and a study of the genetic susceptibility of humans to group A streptococci and to leprosy. The most versatile model system for studies of disease susceptibility is the inbred mouse. Although seven of the chapters deal with the genetics of the resistance of mice to infection, all of them point out general principles and, wherever possible, parallels with appropriate human diseases.

Genetic studies of the mechanisms of resistance and pathogenesis have an advantage over other approaches. By utilizing animals of appropriate genotypes, it is possible to study the in vivo consequences of variations in particular host defenses in intact animals. Some of the modern genetic approaches used in mouse genetics are also described.

All of the chapters dealing with mouse genetics describe studies with recombinant inbred mice. A chapter has been included that describes approaches for the use of mice in genetic studies of disease resistance. This chapter also describes recombinant inbred mice and a means by which they can be used to examine the linkage of genes affecting disease resistance.

The bulk of the volume focuses on the genetic regulation by three different murine loci: *lps*, *xid*, and *Ity*. *Ity* was the first name given to a locus that governs resistance to infections with leishmania, bacille Calmette-Guérin and salmonella. Different reviews describe the relationship of the *Ity* locus to the diseases caused by two of these three pathogens. These two reviews also include a discussion of other genetic factors that affect the susceptibility of mice to each pathogen. Two reviews are devoted to the X-linked immunodeficiency (*xid*) locus. One describes the present state of immunological knowledge about the immunological deficits that have been shown to be associated with the defective *xid* allele. The other describes the effect of these immunodeficiencies on the susceptibility of mice to infection with *Streptococcus pneumoniae*. Another chapter describes the effects of the *lps* locus on the immune system and the concomitant effects these changes have on the resistance to bacterial infection. There is also a chapter describing genetic studies that examine the relationship between the genetic control of certain macrophage properties and the susceptibility to *Listeria* infection.

Many of the questions concerning the mechanism of action of mammalian resistance genes can be effectively probed using bacteria that have been genetically modified so that they lack one or more virulence properties. It seems likely that much of the future progress using animal systems will rely on genetic manipulations of both the pathogen and the host. The final review describes some of the genetic modifications that have been made in bacterial genes important for the virulence of bacteria in mammals.

Spring 1986   DAVID E. BRILES

Table of Contents

Indexed in Current Contents

List of Contributors

BENJAMIN, Jr., W.H., Cellular Immunobiology Unit of the Tumor Institute, Department of Microbiology, and the Comprehensive Cancer Center, University of Alabama at Birmingham, Birmingham, Alabama 35294, USA

BOOKER, C.L., Cellular Immunobiology Unit of the Tumor Institute, Department of Microbiology, and the Comprehensive Cancer Center, University of Alabama at Birmingham, Birmingham, Alabama 35294, USA

BRILES, D.E., Cellular Immunobiology Unit of the Tumor Institute, Department of Microbiology and Pediatrics, and the Comprehensive Cancer Center, University of Alabama at Birmingham, Birmingham, Alabama 35294, USA

CLAFLIN, J.L., Department of Microbiology and Immunology, University of Michigan Medical School, Ann Arbor, Michigan 48109, USA

COLWELL, D.E., Department of Microbiology, University of Alabama at Birmingham, Birmingham, Alabama 35294, USA

FORMAN, C., Cellular Immunobiology Unit of the Tumor Institute, Department of Microbiology, and the Comprehensive Cancer Center, University of Alabama at Birmingham, Birmingham, Alabama 35294, USA

GIBOFSKY, A., Cornell University Medical College, New York, USA

HOROWITZ, J., Cellular Immunobiology Unit of the Tumor Institute, Department of Microbiology, and the Comprehensive Cancer Center, University of Alabama at Birmingham, Birmingham, Alabama 35294, USA, and University Center for Health Sciences, Bengrion University of Negev, P.O. Box 653, Beer-Sheba, Israel

HUSTER, W.J., Cellular Immunobiology Unit of the Tumor Institute, Department of Biomathematics and Biostatistics, and the Comprehensive Cancer Center, University of Alabama at Birmingham, Birmingham, Alabama 35294, USA

KONGSHAVN, P.A.L., Department of Physiology, 3655 Drummond Street, Montreal, Quebec, H3G 1Y6, Canada

MÄKELÄ, P.H., National Public Health Institute, Mannerheimintie 166, SF-00280 Helsinki, Finland

MCDANIEL, L.S., Cellular Immunobiology Unit of the Tumor Institute, Department of Microbiology, and the Comprehensive Cancer Center, University of Alabama at Birmingham, Birmingham, Alabama 35294, USA

MCGHEE, J.R., Department of Microbiology, University of Alabama at Birmingham, Birmingham, Alabama 35294, USA

MICHALEK, S.M., Department of Microbiology, University of Alabama at Birmingham, Birmingham, Alabama 35294, USA

O'BRIEN, A.D., Department of Microbiology, Uniformed Services University of the Health Sciences, 4301 Jones Bridge Road, Bethesda, Maryland 20814, USA

POSEY, B., Cellular Immunobiology Unit of the Tumor Institute, Department of Microbiology, and the Comprehensive Cancer Center, University of Alabama at Birmingham, Birmingham, Alabama 35294, USA

SCHER, I., Merck Sharp and Dohme Research Laboratories, Rahway, New Jersey 07065, USA

SCOTT, G., Cellular Immunobiology Unit of the Tumor Institute, Department of Microbiology, and the Comprehensive Cancer Center, University of Alabama at Birmingham, Birmingham, Alabama 35294, USA

SKAMENE, E., Division of Clinical Immunology and Allergy, Montreal General Hospital, 1650 Cedar Avenue, Montreal, Quebec, H3G 1A4, Canada

STOCKER, B.A.D., Department of Medical Microbiology, Stanford University School of Medicine, Stanford, California 94304, USA

WICKER, L.S., Merck Sharp and Dohme Research Laboratories, Rahway, New Jersey 07065, USA

ZABRISKIE, J.B., Rockefeller University, 1230 York Avenue, New York, NY 10021-6399, USA

Genetic Control of the Susceptibility to Infection with Pathogenic Bacteria

J.B. ZABRISKIE[1] and A. GIBOFSKY[2]

1 Introduction

The elegant early work of BENACERRAF and GERMAIN (1978), McDEVITT (1976), and others (KLEIN 1975; SNELL et al. 1976) has clearly demonstrated the influence of the mouse histocompatibility complex (H-2) on the susceptibility or resistance of different strains of mice to a variety of diseases including bacterial infections. As a result, numerous investigations have resulted in a veritable explosion of knowledge concerning these interrelationships. Many aspects of these murine H-2 genetic influences on these disease associations will be described in other chapters in this book. Our task is to point out those examples in man in which there is a definite or presumed relationship of the human histocompatibility system (HLA) to disease susceptibility and pathogenesis of bacterial disease.

As a result of the complexity of the system and often the lack of the isolation of a definitive infectious agent, there are surprisingly few examples in which we can firmly establish the genetic influence of the major histocompatibility

---

[1] Rockefeller University, 1230 York Avenue, New York, NY 10021-6399, USA
[2] Cornell University Medical College, New York, USA

Current Topics in Microbiology and Immunology, Vol. 124
© Springer-Verlag Berlin·Heidelberg 1986

complex on resistance or susceptibility to bacterial infections in humans. Among the many factors responsible for this relative dearth of knowledge has been the lack of animal models resembling the human disease which would extend our knowledge of the interaction between microbe and host in a given clinical syndrome. In addition, the difficulty in recognizing the evolution of disease in several human generations within a short time span often makes it impossible fully to understand the genetic background of susceptible and resistant individuals. Finally, the prolonged interval between an acute infection and subsequent manifestations of clinical disease also makes it difficult to ascribe a direct relationship between a given microbe and the host's susceptibility to that microbe.

In spite of these drawbacks, certain patterns of genetic influence on resistance or susceptibility to microbial disease in man are beginning to emerge, and these examples will form the main theme of this chapter. In addition, we will include examples where a known genetic marker has been reported in increased frequency in certain disease states in which a microbe is *presumed* (although not conclusively documented) to be the responsible agent.

In order to better understand the microbial host interactions, we will first review the salient features of the major histocompatibility complex in man as we presently understand it. Next, we will describe in general terms the various ways in which microbes may avoid the immune system. Finally, we will discuss those diseases and clinical conditions in which a genetic influence appears to play an important role in the susceptibility or resistance of the host to a given microbe and in what manner the microbe initiates or perpetuates the pathological symptoms in a given disease.

2   Major Histocompatibility System

Five closely linked loci, *A*, *B*, *C*, *D*, and *DR*, were defined by agreement at the Eighth International Histocompatibility Workshop in 1980 (TERASAKI 1980). The products of the *A*, *B*, and *C* series are defined using serological reagents, most often with a lymphocytotoxicity assay. The determinants of the *D* locus are defined by cell-to-cell interaction in the mixed lymphocyte culture.

Initial studies directed toward the development of serological methods for the detection of HLA-D antigens have resulted in the subsequent recognition of yet another antigenic system preferentially expressed on the surface of B-lymphocytes. This B-cell system has extensive biological and chemical homologies with the I-region antigens of the murine histocompatibility system and has also been termed Ia. The human Ia alloantigens were primarily recognized with alloantibodies that developed as a result of immunizations with paternal antigens during pregnancy, or in the sera of renal transplant recipients who become immunized against nonmatching antigens present on the homograft. These human Ia determinants are highly polymorphic, with certain alloantigenic specificities related closely to HLA-D alleles. For this reason, these particular Ia phenotypes were designated HLA-DR, to indicate their close relationship with the determinants of the HLA-*D* locus.

Also relevant to this discussion are the studies emerging from the Ninth International Histocompatibility Workshop (ALBERT et al. 1984), indicating that the gene products of the HLA-D region appear to be highly complex, polymorphic, and not yet fully defined. Each product consists of a noncovalently associated combination of an alpha and beta chain. The alpha and beta chains are substantially different from each other, and there is evidence for at least six alpha chain genes and seven beta chain genes, all in the HLA region. These genes appear to be arranged in subsets corresponding to three distinct products all of which are class II molecules: (1) DR molecules, with homology to murine I-E antigens (one alpha and two or three beta chains); (2) DQ molecules, with homology to murine I-A antigens (two alpha and two beta chains); and (3) DP molecules, intermediate in structure between I-A and I-E (two alpha and two beta chains) which appear not to be serologically defined. The alpha and the beta chain genes of each set of products are significantly more similar to each other than to genes of one of the other allelic series. In addition, other investigators have characterized a fourth set of molecules which exhibit homology with DR molecules by peptide mapping. However, these results are preliminary to date.

The antigens of this system are inherited in classical Mendelian fashion. Unlike those phenotypic characteristics which exhibit dominant and recessive forms (e.g., eye color, ABO type), the HLA antigens are codominant; if a gene has been inherited from either parent, the corresponding HLA antigen will be expressed on the cell surface. Given the number of alleles at each locus, there is a large number of possible phenotypic combinations; hence the enormous immunogenetic heterogeneity of an outbred population. Thus, the finding that the frequency of a particular antigen in a patient group is different from that seen in normals is likely to prompt intense interest in the biological role of this system in the regulation of the immune response and disease susceptibility.

3 Mechanisms of Microbial Host Interactions

In an effort to better understand the genetic influence on microbial host interactions, it is first necessary to understand the mechanisms whereby microbes cause pathological damage to the host. In many instances this involves a direct invasion of the susceptible host and unless checked by the host's defenses or some exogenous help (such as antibiotics), the result is death. The above approach is, if you will, self-destructive and does not permit a continued symbiosis of bacterium and host. In order to avoid this situation, many bacteria have developed a more indirect approach to establishing a pathological foothold which is not self-destructive and permits the continued growth and dissemination of the organism. However, each mechanism is worthy of discussion and we shall illustrate the various pathways utilized by bacteria to gain entrance to the host with specific examples. Whenever possible we will endeavor to show how the host's immune response (or lack thereof) results in a clinical disease process.

These pathways may be conveniently grouped into four major headings:

1. Direct invasion
2. Organisms that exert their effect by microbial mimicry
3. Organisms that establish pathological footholds via suppression of the immune response
4. Organisms that release substances with specific biological properties leading to immunopathological damage

3.1 Direct Invasion

Once again there are numerous animal models in which the response to a given direct microbial infection appears to be under the genetic influence of the major histocompatibility complex. In most cases this direct invasion is usually related to a virulence factor present on the surface of the organism which confers invasive properties to the microbe or to a toxin elaborated by the organism which either paralyzes the immune system or vital organs of the host. Outstanding examples of the importance of a surface antigen in murine infections would include the marked H-2 influence on infections in different strains of mice with pneumococci (see the Chapter by Briles, Horowitz et al.) salmonella (GLYNN 1983), and many others. With respect to salmonella, the genes conferring resistance appear to be polygenic and not directly related to a given H-2 locus. With respect to steptococcal infections in mice there also appear to be differences in susceptibility of various mouse strains to the injection of type specific group A streptococci (Jones, unpublished data) but the exact mechanisms whereby these mouse strains are susceptible or resistant are at present unknown.

In terms of human disease there appears to be little evidence that resistance or susceptibility to the direct actions of toxins or microbial surface markers are under the genetic control of the host. Natural or acquired immunity to the microbe or its toxins appear to play a much more pivotal role than the MHC. However, as in the case of salmonella studies, resistance or susceptibility to a direct invasion of the organism may be polygenic and involve genes outside the MHC.

3.2 Biological Mimicry

The term molecular mimicry was first used by Raymond DAMIAN (1964) to describe his studies on parasitic organisms, but it is now clear that the phenomenon of shared antigenic determinants between host and microbe is quite common and probably occurs more frequently than has been reported. Table 1 is a partial listing of known cross-reactions between bacteria and various tissue determinants. It is our belief that immunological responses to these cross-reactive antigenic determinants occur constantly in man. As will be seen below, perhaps only in the genetically programmed individual does immunologically relevant pathological damage occur.

Table 1. A selection of cross-reactions between microbes and mammalian tissues

Organism	Tissues	Possible disease association
Streptococcus pyogenes	Heart, brain, kidney, etc.	Rheumatic fever
Streptococcus mutans	Heart	?
Klebsiella	HLA-B27, lymphocytes, man	Ankylosing spondylitis
Salmonella	Mouse tissues	Salmonella infection
Pneumococci 50% gram-negative bacteria	Blood group substances	? Susceptibility to infection
Escherichia coli	Colon tissues	Ulcerative colitis
Trypanosoma cruzi	Neuronal tissue heart tissue	Chagas' disease

The group A streptococcus is, in particular, an outstanding example of bio-logical mimicry and is capable of mimicking certain antigenic determinants of nearly all major human organs of the body (ZABRISKIE 1982). Without boring the reader with all the historical and experimental details, KAPLAN'S investiga-tions (1969) and our own work (ZABRISKIE and FREIMER 1966) indicated that certain streptococcal antigens shared similar epitopes with cardiac antigens. Rabbits immunized with whole streptococci or isolated cellular components thereof produced an antibody which was shown by immunofluorescence to bind to human heart sections. A similar antibody was found in the sera of acute rheumatic fever patients, and the staining pattern of these two antibodies appeared to be identical. The localization of the streptococcal antigens in ques-tion has been controversial, but it now appears that there are at least two antigens: one residing in the cell membrane and one closely associated with the M protein moiety.

With respect to the latter, recent data by Fischetti and coworkers (MANJULA and FISCHETTI (1980) have demonstrated that the M protein moiety exhibits significant structural homology with the muscle protein, tropomyosin. Further-more, DALE and BEACHEY (1982) have noted that peptide fragments of the M protein moiety when injected into rabbits elicit the production of antibodies that are not only opsonic in a phagocytosis system but also bind to cardiac tissue.

Concerning other cross-reactions between group A streptococci and human host tissue, it has also been established that these bacterial antigens cross-react with (1) renal antigens, (2) thymic material, (3) skin, and (4) brain antigens. The significance of these cross-reactions to the disease state has been investigated in many laboratories and extensive reviews of these reactions and their signifi-cance may be found elsewhere (RAPAPORT and CHASE 1964; READ and ZABRISKIE 1976, LYAMPERT et al. 1980).

As one might suspect, there are many other examples of cross-reactions between host tissues and microbial organisms. Foremost among these studies has been the elegant work of PERLMANN and associates (1965), who, over a period of years, have shown that *Escherichia coli* contain antigens which cross-react with antigens present in the human intestinal tract. The sera of patients with ulcerative colitis contain antibodies which react with both intestinal colon antigens and antigens from *E. coli*. An extremely important finding by these authors, and one which may have relevance to the general field of microbial mimicry, is that the cross-reactions were observed primarily with *fetal* intestinal antigens (LAGERCRANTZ et al. 1968). These results suggested that the observed cross-reactivity might be related to "buried" antigenic determinants which are preferentially expressed in fetal tissue. Alternatively, these reactions could occur at the level of transplantation antigens (RAPAPORT and CHASE 1964). That these cross-reactions were seen in other inflammatory bowel conditions (TABAGCHALI et al. 1978) does not necessarily detract from the above observations. As previously mentioned, many cross-reactions between microbes and host exist and certainly not all are deleterious to the host. In all probability, only when these cross-reactions occur in a genetically susceptible individual (see discussion of rheumatic fever, below) does the interaction result in clinical disease.

Another intriguing example of microbial host interactions involves the reaction between *Klebsiella* (and other similar organisms) and the HLA-B27 antigen described originally by EBRINGER (1981) and GECZY et al. (1980). Ebringer postulates that antigens of certain strains of *Klebsiella* cross-react with antigens present in individuals bearing the HLA-B27 marker on their cells. In some manner not completely understood, this cross-reaction somehow results in ankylosing spondylitis. He and his colleagues have suggested that the shared antigenicity between the microbial antigen and a specific HLA class I molecule permits persistence of the microbe since the invading organism is regarded as "self" and therefore is not attacked by the immune system. This persistence may then result in inflammation of host tissue followed by disease. An alternate hypothesis has been suggested by Geczy and colleagues of preferential binding of the microbial antigen to cells bearing the B27 marker. This "self + x" results in pathological damage. The importance of disproving or proving these hypotheses mentioned should not be underestimated as experimental data would provide important clues to pathogenesis in a disease process which has the most closely linked interrelationship with HLA antigens.

A final example of microbial host interactions in which the bacterial antigen apparently acts as a carrier of host tissue is demonstrated by the work of FEIZI (1980), who suggests that the mycoplasmal antigen absorbs the erythrocyte I antigen to its surface, thus acting as an adjuvant for presentation of "altered" host antigens to the immune system. Irrespective of the exact mechanism, patients with mycoplasmal pneumonia often develop cold agglutinins which may then cause immunopathological damage such as hemolytic anemia. Although an uncommon occurrence, this mechanism may be operative in other bacterial-host interactions as well.

3.3 Suppression of the Immune Response

The third mechanism by which pathogenic organisms achieve pathological damage is by circumvention or suppression of the host's immune response. During our studies of acute post streptococcal nephritis we noted that the blastogenic response to certain streptococcal antigens in nephritic patients was decreased as compared with controls (REID et al. 1984). We next separated the blastogenic responses to streptococcal antigens by age, given the earlier observation by BALDWIN (1974) that APSGN contracted after the age of 10 years had a much worse prognosis. The data indicated that acute nephritic patients <10 years of age were similar to controls in their response to the streptococcal antigens. In marked contrast, patients over the age of 10 years had an definite antigen-specific cellular suppression to the streptococcal antigens but a normal response to lectins. Suspecting that an adherent cell population might be responsible for the observed suppression, we then tested the reaction to these antigens following removal of those cells. The removal of the adherent cell population markedly enhanced the response of nephritic patients to these streptococcal antigens. Reintroduction of adherent cells back into the assay system promptly depressed the response to the streptococcal antigens.

Numerous investigators have demonstrated that cellular suppression to various specific and nonspecific mitogens occurs in a number of infections caused by bacteria viruses and protozoa. For the purpose of this chapter, we will consider only bacterial indirect cellular suppression in disease states. Among the best studied is the cellular suppression observed in patients with tuberculosis. (KATZ et al. (1979) noted the presence of suppressor monocytes in untreated TB patients, associated with a cellular suppression to both pokeweed mitogen and mycobacterial antigens. Removal of this adherent non-T-cell population resulted in a marked increase in the cellular response. Treatment of these patients with a regimen of antituberculous drugs for 4–6 weeks resulted in clinical improvement as well as a complete return of normal cellular response.

Recently, VAN VOORHIS et al. (1982) have reported that the majority of T cells in the cutaneous lesions of patients with lepromatous leprosy were predominantly OKT_8^+ (suppressor cell); few OKT_4^+ (helper) cells were identified. In contrast, the majority of T cells in the cutaneous lesions of patients with tuberculoid leprosy were OKT_4^+. The authors suggest that OKT_4^+ cells may have a strong influence on the microbicidal activity of the macrophages in the lesions. These studies also indicate that soluble factors secreted by those macrophages laden with the infectious organism may lead to a specific signal for one or another T-cell population. If true, these observations are of some significance since they indicate the microbe itself directs an immune cell to release lymphokines resulting in a tissue tropism for a specific suppressor or helper cell at the cutaneous site of inflammation.

3.4 Direct Immunopathological Damage

In this final mechanism, substances released by the organism appear to have specific biological properties for the host, resulting in immunopathological damage. An example is shown by the elegant work of VILLARREAL and collaborators (1979) which demonstrated that nephritogenic strains of group A streptococcus secrete a protein with a molecular weight of 46 000. Polyclonal antibodies prepared to this protein stain the majority of the renal biopsy specimens obtained from patients with APSGN while nonstreptococcal-induced glomerular biopsies were uniformly negative.

While the presence of this protein in the biopsy specimens argued strongly for a causal association of the protein with the disease process, how the protein localizes in the kidney has been the subject of more recent experiments. Since our previous studies (VAN DE RIJN et al. 1978) demonstrated that patients with APSGN had high levels of circulating immune complexes, the question of whether or not these complexes contained streptococcal antigens was first investigated. Complexes were isolated using polyethylene glycol precipitation followed by column purification. These purified complexes were injected into rabbits and anti-immune complex antibodies produced. This anti-immune complex rabbit serum contained antibodies which reacted with streptococcal extracellular preparations obtained from nephritogenic strains. In contrast, rabbit antisera made against circulating complexes isolated from acute rheumatic fever patients reacted with antigens secreted by rheumatogenic strains; and the patterns were different in both groups (FRIEDMAN et al. 1984).

Thus, extracellular antigens unique to those nephritogenic strains associated with the disease are present in the circulating immune complex and are also deposited in the glomerular tissues of these patients. We think it likely that this antigen has unique biological properties, one of which may be its ability to trigger the alternate pathway of complement. This property alone or in combination with the circulating complex could cause the pathological damage that results in APSGN. This could also explain the well-known observation that streptococcal antigens and complement (including components of the alternative pathway) appear in the glomerular lesions prior to the appearance of specific host antibody in the lesion (SEEGAL et al. 1965).

The list of exogenous bacterial antigens implicated in various forms of nephritis is extensive and has been recently reviewed by VILLARREAL et al. (1982). With many of the other bacterial antigens, a pathological picture of proliferative nephritis similar to that seen in poststreptococcal glomerulonephritis is seen as well. Since many of these bacterial antigens may trigger the alternative pathway of complement activation, the mechanism for inducing renal damage may be similar in *all* postbacterial glomerulonephritides.

Lest the reader the left with the impression that bacterial antigens act only via complement activation or immune complex formation, it has also been proposed that certain organisms secrete sialidases which may alter the composition of host immunoglobulins, causing them to be perceived as "foreign." Complexes of altered immunoglobulin anti-immunoglobulin antibodies would then be formed, eventually resulting in pathological damage. Evidence for this concept is found in the observation that cryoglobulins are present in a large proportion of APSGN patients. Furthermore, certain strains of group A streptococci produce sialidases that are capable of removing sialic acid from IgG molecules (McINTOSH et al. 1975).

4 Disease Associations

Of the many conditions thus far investigated and shown to be associated with particular alleles of the HLA system, the most important have been the rheumatic diseases. While the associations are high, they are neither absolute nor diagnostic. The presence of a specific antigen is not the sole factor in disease pathogenesis, as the antigen occurs in a low but definite frequency in disease-free individuals as well. Nevertheless, knowledge of the association may prove useful in permitting *subdivisions* of clinical groups within the larger population, for example, pauciarticular juvenile rheumatoid arthritis. This could facilitate the search for possible etiological agents and confirm or refute the following suggested mechanisms for HLA and disease associations: (1) the HLA antigen might be structurally similar to the antigenic component of an infectious agent, (2) the HLA antigen may be a part of a neoantigen, formed in combination with an infectious agent, (3) the HLA antigen may be a receptor for an infectious or environmental toxin, (4) there may be a deficiency of one or more complement system components, and (5) there may be linkage disequilibrium with one or more immune response genes. A list of diseases associated with particular alleles of the HLA system follows.

5 Diseases Associated with the MHC

5.1 Rheumatic Fever

The evidence that rheumatic fever patients have an abnormal response to streptococcal antigens can be gleaned from two sources. On the humoral side, the examination of a large number of rheumatic fever sera indicate that patients with the acute disease have elevated titers of antibodies which bind to muscle tissue including cardiac tissue (ZABRISKIE et al. 1970). Absorption studies with this antibody have revealed that all of the tissue staining can be absorbed by a solubilized and highly purified fraction of group A streptococcal membranes as well as by cardiac antigens (VAN DE RIJN et al. 1977). In contrast, heart-reactive antibodies which appear during post pericardiotomy syndrome are absorbed only by cardiac antigens (ENGLE et al. 1984). Patients with established rheumatic valvular disease also contain an antibody which reacts with extracted valvular glycoprotein material and the group-specific carbohydrate of the group A steptococcus (GOLDSTEIN et al. 1968; DUDDING and AYOUB 1968). The shared antigenic determinant between the two molecules is *N*-acetyl glucosamine. Finally, there is an antibody seen in rheumatic fever chorea patients which binds to the cytoplasm of caudate nuclei of the brain and can be absorbed by streptococcal membrane antigens (HUSBY et al. 1976).

On the cellular side, there is an increased cellular response of these patients to streptococcal antigens (in particular, membrane antigens isolated from streptococcal strains commonly associated with rheumatic fever) that persists for at least 2 years after the initial attack (READ et al. 1974). In addition, these patients also respond simultaneously to extracts of cardiac tissue (GOWRISHAN-KAR and AGARWAL 1980). The question of whether or not these mononuclear populations are indeed specifically cytotoxic for heart cells has been difficult to resolve. Experimental evidence (YANG et al. 1977) would indicate that animals sensitized to streptococcal membranes (but not cell walls) produce lymphocytes which are cytotoxic for embryonic cardiac tissue monolayers. Of interest was their observation that the addition of heart-reactive antibody to these cell cultures did not enhance the cytotoxic effects. Only one report in humans (HUTTO and AYOUB 1980) indicates that the lymphocytes obtained from a few acute rheumatic fever patients are cytotoxic for human cardiac tissue, but the number of patients studied was small and heart-reactive antibody from these patients was not tested in the cytoxicity system. However, in the Aschoff lesions themselves we do see evidence of T-cell subset imbalances, suggesting that cellular mechanisms do play an important role in the pathology of rheumatic heart disease (RAIZADA et al. 1983).

If, as was suggested at the outset, these microbial host cross-reactions occur in many individuals, why do only a few individuals with streptococcal infections go on to develop rheumatic fever? In fact, starting with CHEADLE (1889) in his Harvean lectures, it has been everyone's suspicion that rheumatic-fever-susceptible individuals were in some way genetically predisposed to contract the disease. However, numerous attempts to identify a genetic marker associated with the disease were unsuccessful or controversial (MURRAY et al. 1982). More recently with the discovery of the close association of the major histocompatability complex and immune responses genes, a further unsuccessful search was made to link the frequency or absence of an HLA antigen of the ABC loci to the disease complex (FALK et al. 1973).

It was not until 1979 when investigators at Rockefeller University in collaboration with Dr. Patarroyo in Colombia, South America (PATARROYO et al. 1979), made the interesting observation that a B-cell alloantigen was found in these patients that appeared to be closely associated with an increased susceptibility to rheumatic fever. In fact, the relative risk factor of contracting rheumatic fever is 12 times greater for individuals with this marker than for those individuals who do not exhibit the marker. Of potential public health importance was the observation that the antigen was present in approximately the same number of rheumatic fever individuals whether they were identified in New York or Bogota, suggesting a worldwide distribution of the antigen.

One of the most fascinating properties of this particular B-cell alloantigen (called 883) is that it does not appear to be related to any of the known HLA, *A*, *B*, *C*, and *DR* loci now known. However, it should be emphasized that at the time of the testing with the original human alloantisera, not all alleles of the *DR* locus were completely defined and, as mentioned above, new subdivisions of the *DR* locus are being discovered. Preliminary data by WINCHESTER and KUNKEL (1979) suggested the possibility that the antigen might be an Ia molecule but further biochemical definition of this antigen has not yet been fully accomplished. Unfortunately, the supply of the original human alloantisera has been exhausted and attempts to define an alloreagent of equivalent specificity have proven unrewarding, despite extensive screenings.

In an effort to confirm and extend the original results, two hybridoma clones were produced which identified a B-cell marker present in approximately 95% of rheumatic individuals. The first clone was almost identical to the original 883 alloantiserum, while the second clone identified the majority of rheumatic fever patients originally labeled 883⁻. Preliminary data with these monoclonals would suggest that these B-cell antigens may not be "classical" Ia antigens but rather a newly defined set of B-cell antigens.

The availability of these monoclonal reagents is of some importance as it may now be possible to identify those individuals at greatest risk *prior* to the onset of the disease, which would with appropriate treatment of the streptococcal infection hopefully prevent the disease. In addition, should streptococcal vaccines by developed, the high-risk group would be easily identified as the most likely candidates for initial use of the vaccine.

Perhaps more importantly these clones may provide clues as to the nature of the pathogenetic mechanisms involved in the rheumatic fever attack. For example, is the antigen present on buccal or tonsillar cells, the crucial point of entry for streptococci that eventually leads to rheumatic fever? Secondly, could these alloantigens be receptors for streptococcal antigens? Thirdly, in what manner does the presence of this antigen on a B-cell regulate the immune response to streptococcal antigens? Finally, how does the antigen segregate in the families of rheumatic fever patients? Answers to all of these important questions could provide a clearer picture of the relationship of this B-cell marker to the actual disease process.

5.2 Leprosy

Although numerous individuals in endemic areas may become infected with *Mycobacterium leprae* and even develop an immune response to the organism, few actually develop any clinical manifestations of leprosy (VAN EDEN and DE VRIES 1984). However, once the disease is established in a given individual, two forms of the disease are apparent. Certain individuals develop a tuberculoid (or TT) leprosy, in which the lesions are granulomatous in nature with few bacteria present within the lesion. The cellular response to the organism is excellent in these patients and the T_4 helper lymphocyte predominates in and around the granulomas. The second form, lepromatous (or LL) leprosy, is more ominous. Numerous bacteria are present within the lesions and in the surrounding macrophages, reflecting a state of relative cellular energy. The predominant cell in these lesions is a T_8 suppressor cell (GODAL 1978). Whether this represents a genetic defect on the part of the host in its inability to generate T-helper cells or whether the organism in situ in some manner stimulates the production of suppressor cells (thereby enabling the organism to persist in the tissues) is an area of active investigation.

The classification of leprosy into two distinct clinical forms of the disease has prompted a great deal of research into the possibility that certain MHC markers might be involved in determining which form the disease might take in a given infected individual. At the present time, several associations have been reported; however, these reports are preliminary and require further confirmation. It would appear that there is an increased frequency of the HLA-DR2-associated alloantigen DQw1 with lepromatous leprosy (LL), while DR3 seems to be closely associated with the tuberculoid form (TT) of the disease (VAN

EDEN et al. 1985). In addition, family studies of leprosy patients suggest that while no one HLA antigen is associated with susceptibility/resistance to leprosy per se, the HLA complex does appear to influence which clinical form the disease will take once an individual is infected with the organism. The present data are consistent with the hypothesis that predisposition to leprosy may be via an autosomal dominant manner of inheritance while predisposition to TT leprosy may be via an autosomal recessive manner of inheritance (VAN EDEN and DEVRIES 1984).

Given the two different clinical forms of leprosy, the data supporting the above hypothesis would suggest that these HLA-linked susceptibility genes may be operating at the level of the immune response to the invading organism. Thus, once infected, the individual who goes on to develop tuberculosis leprosy will mount an intense cellular and humoral response to the organism. In contrast, certain infected individuals will develop lepromatous leprosy as they will be unable to mount an effective immune response to the organism. This may occur as a result of a defective T-cell response, or an inability to recognize those *M. leprae* bacterial antigens which may closely resemble self-antigens. This lack of response to the organism in LL patients may be further augmented by factors released by the organism itself which cause the host to produce certain lymphokines which enhance cellular suppression. Similar observations have been reported in patients with poststreptococcal glomerulonephritis and tuberculosis (see previous section on streptococcal antigens).

Further knowledge of the HLA linkage to these two different forms of leprosy may have important public health considerations. For example, as VAN EDEN and his collegues (1984) have pointed out, the existence of a suppressor gene in leprosy (I_s) in those individuals who eventually develop LL forms of leprosy might prevent those individuals from responding to any proposed *M. leprae* vaccines.

5.3 Tuberculosis

Of the group of granulomatous diseases discussed above, tuberculosis (TB) still ranks as one of the most common of the infectious diseases. Estimates of 8 million new cases/year are not unreasonable (STYBLO 1983) and may be higher, as many cases probably go undetected in third world countries. Factors such as poor nutrition, overcrowding, and poor health care delivery which affect other infectious diseases are also important in the spread of this disease. These conditions coupled with the apparent failure of bacille Calmette-Guérin (BGG) to protect against TB make this one of the most challenging problems in modern medicine.

The question of whether or not there is an MHC marker associated with resistance or susceptibility to TB has been examined on several occasions (see review by POLLACK and RICH 1985) and the data to date indicate that there is a suggestion of an HLA-DR (probably DR2 related)-associated susceptibility factor to the disease. Perhaps the most interesting study has been carried out by PATARROYO (1980) in which a single DR2 alloantiserum detected a marker present in 75% of TB patients. Fourteen other DR2 alloantisera were negative, suggesting that the marker is closely associated with DR2 but not identical to it.

Questions as to whether susceptibility to TB in man is related to a defective immune response to the organism on the part of the host or represents immuno-suppression of the host's defenses by the organism remain unanswered. It is our belief that many of the mechanisms which appear to be operative in leprosy may well be applicable in future studies of TB.

5.4 Lyme Disease

A peculiar clinical syndrome characterized by recurrent episodes of acute syno-vitis, primarily in large joints, this entity is now recognized to have multisystemic manifestations which may occur long after the first arthropathic manifestations. The disease was recognized a decade ago by Steere and Malawista (STEERE et al. 1977), who described a geographical clustering of cases in the communities of Lyme, Old Lyme, and East Haddam, along the Connecticut River.

Numerous epidemiological studies suggested that this was most likely a chronic infectious illness caused by an agent transmitted by the bite of the deer tick, *Ixodes dammini*, an arthropod vector with endemic habitation in those areas (wooded) in which the disease was described. Recent work by BURGDORFER and his colleagues have demonstrated that the agent responsible is a spirochete (BURGDORFER et al. 1982), carried by the tick, which is then injected into the skin or blood of the individual, who then goes on to develop the disease. Isola-tion or direct identification of the spirochete allows for a definitive diagnosis, but this is quite rare. More commonly, the diagnosis is made by measuring antibody titers against the spirochete. As Steere has reported (STEERE 1979), specific IgM antibody titers usually peak between 3 and 6 weeks postinfection, while specific IgG titers rise more slowly and are generally increased several months later. These antibody titers may be useful in differentiating lyme disease from other rheumatic syndromes.

Lyme disease is of particular interest in that it is one of the few rheumatic diseases in which an infectious agent has been conclusively established. In addi-tion, there is also good evidence for disease susceptibility, mediated through genes of the major histocompatibility complexes. An increased frequency of the B-cell alloantigen HLA-DR2 is seen in patients with the severe chronic form of the illness. While exact pathogenic mechanisms reponsible for the myr-iad of symptoms are not known, there is a strong suggestion of a microbial-host interaction. Sera of patients with the disease do exhibit evidence of heart-binding antibodies (Zabriskie, unpublished data) and there are well-known serological cross-reactions between numbers of the spirochetal family and cardiolipin, a mammalian host protein found in many organs of the body including heart tissue (WILKINSON 1984). Thus the elements are in place for another cross-reaction between host tissues and microbial antigens similar to that proposed for rheumatic fever pathogenesis. Hopefully future investigations in the area of microbial host interactions in this disease will provide answers concerning the initiation of the various symptoms of Lyme disease.

5.5 Major Histocompatibility Complex-Related Seronegative Arthritis

Perhaps the best known of the clinical enthesiopathies, anklyosing spondylitis (AS), is an entity characterized by progressive inflammation and ultimate fusion of the sacroiliac joints and axial skeleton. Although the primary target tissue is fibrocartilage, extra-articular manifestations (e.g., anterior uveitis, aortitis) may also occur. Of particular interest, a significant increased frequency (approximately 90%) of the class I antigen HLA-B27 has been reported in white patients with this disease (BREWERTON et al. 1973; SCHLOSSTEIN et al. 1973).

The exact pathogenesis of AS is not known; however, several lines of evidence seem to implicate an infection by a particular agent as the initiating event. Several investigators have postulated that certain strains of bacteria (e.g., *Klebsiella, Shigella* and *Yersinia*) carry plasmids which code for factors modifying HLA-B27, eliciting autoimmune responsiveness and chronic infection (GECZY et al. 1980; EBRINGER 1981). Other investigators have reported cross-reactivity between bacterial surface antigen and B27 (KONO et al. 1985). Much additional work remains before either hypothesis can be conclusively established or refuted; however, the recognition of significant clinical arthropathy in the B27-positive individual following enteric infection lends support to the infections etiology of this chronic rheumatic disorder and further supports the interrelationship between genetic susceptibility and environment in disease pathogenesis.

While the HLA-B27 marker is closely associated with the disease in some populations, there are many well-documented cases of AS which are B27 negative. Data of this nature raise the question of whether there are several genes involved in the pathogenesis of the disease or alternatively whether the gene is associated with HLA-B27 marker but not identical to it.

The second clinical syndrome associated with the HLA-B27 is Reiter's syndrome, which also has a high predeliction for young males. The disease usually occurs one to several weeks after a diarrheal illness or in some patients, following extramarital intercourse; and the classic "triad" of clinical manifestations are arthritis, conjunctivitis, and urethritis (PARKER 1980).

A variety of enteritis-causing microorganisms (e.g., *Shigella, Yersinia, Salmonella, Amoeba*) have been implicated as causative agents in the syndrome. In this context, one of the most interesting "experiments of nature" occurred. Certain members of a crew (602/1276) on board a naval vessel in 1962 contracted *Shigella* bacillary dysentery. Ten cases of Reiter's syndrome occurred among those 602 crewmen (NOER 1966). Subsequent follow-up of five of the original ten patients 13 years later revealed that four out of the five individuals were all HLA-B27 positive and had persistent disease. The remaining affected individual was HLA-B27 yet negative and had minimal disease (CALIN and FRIES 1976).

The frequency of HLA-B27 in patients with Reiter's syndrome is approximately 70% in most series (4–8% in control populations). However, when true spondylitis or sacroileitis is a feature, the likelihood of finding B27 in the affected individual is higher. As in ankylosing spondylitis, the same questions concerning the relationship of the microbe to a given HLA antigen are also applicable to Reiter's syndrome.

5.6 Complement Disorders

It has been known for 100 years that the killing of certain infectious organisms required both a serum heat-stable factor (antibody) and a serum heat-labile factor (complement). Extensive study of this complement system has revealed that the various components are important in numerous facets of the immune response including chemotaxis, opsonization, phagocytosis, and certain cyto-toxic cellular defense mechanisms. The complement system can be activated by both gram-positive and gram-negative organisms and thus deficiencies in various complement components theoretically could account for susceptibility to infectious diseases.

In spite of this enormous theoretical susceptibility to infections in those individuals deficient in one or more complement components, clinically signifi-cant infection in humans deficient in one or more of the complement components is unusual. For example, patients deficient in C1, C2, and C4 are *not* unduly susceptible to infectious diseases, and patients deficient in terminal components (C5 through C8) appear to be susceptible largely only to gram-negative *Neisseria* infections. Individuals deficient in C3 are susceptible to repeated bacterial infec-tions and are clinically similar to individuals with hypogammaglobulinemia (AT-KINSON and FRANK 1980).

Several studies have clearly demonstrated the close association between the genes of the MHC complex and the genes coding for several complement compo-nents (LACHMANN and HOBART 1978) particularly C4 and C2 and possibly C3, C6, and C8. In contrast, the genes coding for C1, C5, C7, and C9 do not appear to be associated with genes of the MHC complex. The linkage between C2 deficiency and certain HLA antigens has been most carefully studied (FU et al. 1974; AGNELLO 1978). Individuals deficient in C2 have a "lupus-like" illness and have an increased frequency of the HLA haplotypes A10-B18-Dw2. Insofar as many apparently healthy individuals in these family units have inher-ited this haplotype as well, other genetic and possibly environmental factors may be relevant to developing clinical disease.

In summary, there appears to be excellent evidence for the association of genes coding for certain complement components with genes of the HLA com-plex. There exist few data, however, in support of the concept that these comple-ment component deficiencies (with the rare exceptions discussed above) result in susceptibility or resistance to disease.

6 Presumed Microbial Associations

In this group of disorders are a number of disease states in which the relationship of the MHC to the clinical state appears to be quite clear but there are few or no actual data concerning a presumed inciting organism (see Table 2). High on the list of a possible infectious etiology to explain these disease states would be rheumatoid arthritis in which the suspicion that a microbe might be involved in the initiation and/or perpetuation of the disease has existed for some time.

Table 2. Major histocompatibility complex associations in autoimmune diseases and possible infectious agents

Disease	HLA association	Possible infectious agents
Rheumatoid arthritis	DR4	Epstein-Barr virus, bacterial peptidoglycans
Lupus erythematosus	DR2, DR3, DQw1	Type C virus
Multiple sclerosis	DR2	Measles virus, canine distemper virus

This concept has been strengthened by a series of elegant experiments by Cromartie and Schwab (CROMARTIE et al. 1977) in which they developed a experimental model of rheumatoid arthritis in rats following a single dose of streptococcal cell walls which mimics many of the features of the human disease. More recent experiment have shown that a variety of microorganisms all of which have the common feature of containing peptidoglycans with rhamnose side chains within their cell wall structure can elicit the experimental relapsing arthritis (CROMARTIE 1981).

With respect to human rheumatoid arthritis there are several reports which claim that antibodies to bacterial peptidoglycans are elevated in rheumatoid arthritis patients especially in juvenile patients with this disease. In addition, a possible streptococcal etiology to systemic juvenile rheumatoid arthritis is suggested by the fact that approximately 40% of these patients have elevated Anti Streptolysin "O" (ASO) titers at the onset of their disease. However, it should be emphasized that these elevated serological reactions in rheumatoid arthritis patients may merely reflect the possible B-cell polyclonal activation so often seen in these patients (HEYMER et al. 1976).

While a bacterial etiology for rheumatoid arthritis in man is both attractive and plausible in view of the wide variety of organisms which can elicit the experimental arthritis, the evidence demonstrating the presence of microbes or their antigens in human joint tissue is sparse. However, the pathological reaction may be similar to that seen in rheumatic fever in which the microbe has stimulated an autoimmune response to host tissue long after the actual microbe has disappeared.

Another area of continuing interest is the relationship of psoriasis to a antecedent microbial infection. Evidence by GROSS et al. (1980) indicated that patients with guttate psoriasis have a heightened cellular reactivity to streptococcal cell walls and membranes. In general, psoriasis patients also have an increased frequency of the HLA antigens B13, B17 and the subgroup with psoriatic arthritis have an increased frequency of BW38. However, in terms of the increased cellular response to streptococcal antigens there was inverse correlation between the presence of the B13, B17 antigens and a heightened response to the streptococcal antigens. In view of the known cross-reactions between fibroblasts and endothelial cells and streptococcal antigens (KINGSTON and GLYNN 1971) the possibility exists that cellular or humoral reactivity to streptococcal antigens cross-reactive with skin might be involved in the disease process.

Finally, there is a general group of autoimmune disorders which include lupus erythematosus, multiple sclerosis, and others in which no definite microbe has been implicated; yet, one suspects that there is an association with a microorganism in these disease states given the known increased frequency of a given HLA antigen in these patients. Future research should be directed toward looking for evidence of "footprints" left behind by these presumed organisms. Hopefully, these studies will shed future light on the pathogenic mechanisms involved in this group of diseases and increase our knowledge concerning the subtle interplays between microbe and host which eventually results in a disease process.

References

Agnello V (1978) Complement deficiency states. Medicine 57:1–23

Albert ED, Baur MP, Mayr WR (eds) (1984) Histocompatibility testing 1984. Springer, Berlin Heidelberg New York Tokyo

Atkinson JP, Frank MM (1980) Complement. In: Parker CW (ed) Clinical immunology, vol I. Saunders, Philadelphia, pp 219–271

Baldwin DS, Gluck MD, Schacht RG et al. (1974) The long term course of poststreptococcal glomerulonephritis. Ann Intern Med 80:342

Benacerraf B, Germain RN (1978) The immune response genes of the major histocompatibility complex. Immunol Rev 38:70–119

Brewerton DA, Hart FD, Nicholls A (1973) Ankylosing spondylitis and HLA W-27. Lancet 1:904–907

Burgdorfer W, Barbour AG, Hayes SF, Benach JL, Grunwaldt E, Davis JP (1982) Lyme disease – a tick bourne spirochetosis? Science 216:1317–1319

Calin A, Fries JF (1976) An "epidemic" of Reiter's syndrome revisited. Followup evidence on genetic and environmental factors. Ann Intern Med 84:564–566

Cheadle WB (1889) Harveian lectures on the various manifestations of the rheumatic state as exemplified in childhood and early life. Lancet I:821, 871, 921

Cromartie WJ (1981) Arthropathic properties of peptidoglycan-polysaccharide complexes of microbial origin. In: Deicher W, Schutz LC (eds) Arthritis Models and mechanisms. Springer, Berlin Heidelberg New York, pp 24–38

Cromartie WJ, Craddock JG, Schwab JH, Anderle SK, Yang CH (1977) Arthritis in rats after systemic infection of streptococcal cells or cell walls. J Exp Med 146:1585–1602

Dale JB, Beachey EH (1982) Protective antigenic determinant of streptococcal M proteins sheared with sarcolemmal membrane protein of human heart. J Exp Med 156:1165–1176

Damian RT (1964) Molecular mimicry: antigen sharing by parasite and host and its consequences. Am Naturalist 98:129–149

Dudding BA, Ayoub EM (1968) Persistance of streptococcal group A antibody in patients with rheumatic valvular disease. J Exp Med 128:1081

Ebringer RW (1981) HLA-B27 and the link with rheumatic diseases: recent developments. Clin Sci 59:405–410

Engle MA, Gay WA, Zabriskie JB, Senterfit LB (1984) The post pericardiotory syndrome: 25 years experience. J Cardiovas Med 9:321–332

Falk JA, Fleishman JL, Zabriskie JB, Falk RE (1973) A study of HL-A antigen phenotype in rheumatic fever and rheumatic heart disease patients. Tissue Antigens 3:173–178

Feizi T (1980) Monoclonal antibodies of cold agglutinin syndrome: immunochemistry of biological aspects of the target antigen with special reference to the large I small I antigen. Med Biol 58:123–127

Friedman J, van de Rijn I, Ohkuni H, Fischetti VA, Zabriskie JB (1984) Immunological studies of post-streptococcal sequelae. Evidence for presence of streptococcal antigens in circulating immune complexes. J Clin Invest 74:1027–1034

Fu SM, Kunkel HG, Dupont B, Hansen JA, Day NK, Good RA, Jersild C, Fotino M (1974) Mixed lymphocyte culture determinants and C_2 deficiency: LD 7A associated with C_2 deficiency in four families. J Exp Med 142:495–506

Geczy AF, Alexander K, Bashir HV, Edmonds I (1980) A factor(s) in *Klebsiella* culture filtrates specifically modifies an HLA-B27 associated cell surface complement. Nature 283:782–784

Glynn AA (1983) Resistance to infection: looking for new genes. Bull Eur Respir 19(2):143–5

Godal T (1978) Immunological aspects of leprosy-present status. Prog Allergy 25:211–42

Goldstein I, Rebeyotte P, Parlebas J, Halpern B (1968) Isolation from heart valves of glycopeptides which share immunological properties with *Streptococcus haemolyticus* group A polysaccharides. Nature 219:866–868

Gowrishankar R, Agarwal SC (1980) Leucocyte migration inhibition with human heart valves glycoproteins and group A streptococcal ribonucleic acid proteins in rheumatic heart disease and post streptococcal glomerulonephritis. Clin exp Immunol 39:519–525

Gross WL, Pachhauser U, Hahn G, Poschmann A (1980) Lymphocyte response to streptococcal antigens and humoral autoimmune phenomena as additional markers of HLA-defined subgroups of psoriasis. In: Read SE, Zabriskie JB (eds) Streptococcal diseases and the immune response. Academic, New York, pp 377–389

Heymer B, Schliefer KH, Read SE, Zabriskie JB, Krause RM (1976) Detection of antibodies to bacterial cell wall peptidoglycan in human sera. J Immunol 117:23–26

Husby G et al. (1976) Antibodies reacting with cytoplasm of subthalamic and caudate nuclei neurons in chorea and acute rheumatic fever. J Exp Med 144:1094–1110

Hutto JH, Ayoub EM (1980) Cytotoxicity of lymphocytes from patients with rheumatic carditis to cardiac cells in vitro. In: Read SE, Zabriskie JB (eds) Streptococcal diseases and the immune response. Academic, New York, pp 733–738

Kaplan MH, Frengley JD (1969) Autoimmunity to the heart in cardiac disease. Current concepts of the relation of autoimmunity to rheumatic fever, postcardiotomy and postinfarction syndromes and cardiomyopathies. Am J Cardiol 24:459

Katz P, Goldstein R, Fauci A (1979) Immunoregulation in infections caused by *Myocobacterium* tuberculosis: the presence of suppressor monocytes and the activation of subpopulations of T lymphocytes. J Infect Dis 18 140:12–21

Kingston D, Glynn LE (1971) A cross reaction between *Streptococcus* pyogenes and human fibroblasts, endothelial cells and astrocytes. Immunology 21:1003

Klein J (1975) Genetic polymorphism of the histocompatibility-2 loci of the mouse. Annu Rev Genet 8:63–78

Kono DH, Ogasawara M, Prince H, Park M, Toivanen S, Yu DTY (1985) Mimicry between HLA-B27 and bacteria: assessment of the probability and the possible clinical significance. Fed Proc 44:597

Lachman PJ, Hobart MJ (1978) Complement genetics in relation to HLA. Br Med Bull 34:232–247

Lagercrantz R, Hammarstrom S, Perlmann P, et al. (1968) Immunological studies in ulcerative colitis IV origin of autoantibodies. J Exp Med 128:1339–1352

Lyampert IM, Kolesnikova VJu, Gnezditskaya EV, Danilova TA, Belctskaya LV, Borodiyuk NA (1980) Cross reactive antigens in group A streptococci: their role in the autoimmune process in pathogeneic streptococci. In: Parker MT (ed) Pathogenic streptococci. Reed Books, Survey, pp 105–106

Manjula BN, Fischetti VA (1980) Tropomyosin-like seven residue periodicity in three immunologically distinct streptococci M proteins and its implications for the anti-phagocytic property of the molecule. J Exp Med 151:695–708

McDevitt HO (1976) The evolution of genes in the major histocompatibility complex. Fed Proc 35:2168–2173

McIntosh RM, Griswold WR, Chernack WB, et al. (1975) Cryoglobulins III. Further studies on the nature, incidence, clinical, diagnostic, prognostic and immunopathologic significance of cryoproteins in renal disease. OJ Med 44:85

Murray GC, Montiel MM, Persellin RH (1978) A study of HLA antigens in adults with acute rheumatic fever. Arthritis Rheum 21(6):652–656

Noer HR (1966) An "experimental" epidemic or Reiter's syndrome. JAMA 197:693–698

Parker CW (1980) Seronegative HLA-related arthritis. In: Parker CW (ed) Clinical Immunology, vol II. Saunders, Philadelphia, p 753

Patarroyo ME (1980) Identification de marcadores geneticos asociados a tres enfermedades comunicables: Lepra, TBC, Fievre Rheumatica. Acta Med Colombiana 5:235

Patarroyo ME, Winchester R, Vejerano A, Gibofsky A, Chalem F, Zabriskie JB, Kunkel HG (1979) Association of B-cell alloantigen with susceptibility to rheumatic fever. Nature 278:173–174

Perlmann P, Hammerstrom S, Lagercrantz R, et al. (1965) Antigen from colon of germ free rats and antibodies in human ulcerative colitis. Ann NY Acad Sci 124:377–394

Pollack MS, Rich RR (1985) The HLA complex and pathogenesis of infectious diseases. J Infect Dis 151:1–8

Raizada V, Williams R, Jr, Chopra P, Copinath N (1983) Tissue distribution of lymphocytes in rheumatic heart valves as defined by monoclonal anti T cell antibodies. AMJ Med 74:90–96

Rapaport FT, Chase RM, Jr, (1964) Homograft sensitivity induction by Group A streptococci. Science 145:407–410

Read SE, Zabriskie JB (1976) Immunological concepts in rheumatic fever pathogenesis. In: Misher PA, Muller-Eberhard HJ (eds) immunopathology. Grune and Stratton, New York, pp 471–487

Read SE, Zabriskie JB, Fischetti VA, Utermohlen V, Falk R (1974) Cellular reactivity studies to streptococcal antigens in patients with streptococcal infections and their sequelae. J Clin Invest 54:439–450

Reid HMF, Read SE, Zabriskie JB, Ramkissoon R, Poon-King T (1984) Suppression of cellular reactivity to group A streptococcal antigens in patients with acute post streptococcal nephritis. J Infect Dis 149:841–850

Schlosstein L, Terasaki PI, Bluestone R (1973) High association of HL-A antigen W27 ankylosing spondylitis. NEJM 288:704

Seegal BC, Andres GA, Hsu KC, Zabriskie JB (1965) Studies on the pathogenesis and acute and progressive glomerulonephritis in man by immunofluorescein and immunoferritin techniques. Fed Proc 24(1):100–108

Snell GD, Dausett J, Nathanson GC (1976) Histocompatibility. Academic, New York

Steere AD (1979) Lyme arthritis: correlation of serum and cyroglobulin IgG with remission. Arthritis rheum. 22:471–483

Steere AC, Malawista SE, Hardin JA, Ruddy S, Ackenase PW, Andiman WA (1977) Erythema chronicum migrans and lyme arthritis: the enlarging clinical spectrum. Ann Intern Med 86:685–698

Styblo K (1983) The epidemiological situation of tuberculosis and the impact of control. Bull Int Union Tuberc 58:179–186

Tabagchali S, O'Donoghue DP, Bettelheim KA (1978) *Escherichia coli* antibodies in patients with inflammatory bowel disease. Gut 19:108–113

Terasaki PI (ed) (1980) Nomenclature for factors of the HLA system. Munksgaard Copenhagen

van de Rijn I, Zabriskie JB, McCarty M (1977) Group A streptococcal antigens cross-reactive with myocardium: purification of heart reactive antibody and isolation and characterization of the streptococcal antigen. J Exp Med 146:579–599

van de Rijn I, Fillit HM, Reid H, Poon-King T, McCarty M, Day NK, Zabriskie JB (1978) Serial studies on circulating immune complexes in poststreptococcal sequelae. Clin Exp Immunol 34:318–325

Van Eden W, DeVries RRP (1984) Occasional review: HLA and leprosy: a re-evaluation. Lepr Rev 55:89–104

Van Eden W, Gonzalez NM, DeVries RRP, Convit J, Van Rood JJ (1985) HLA-linked control of predisposition to lepromatous leprosy. J Infect Dis 151:9–14

Van Voorhis WC, Kaplan G, Sarno EM (1982) The cutaneous infiltrates of leprosy: cellular characteristics and predominant T-cell phenotypes. N Engl J Med 307:1593–1597

Villarreal H, Jr, Fischetti VA, van de Rijn I, Zabriskie JB (1979) The occurrence of a protein in the extracellular products of streptococci isolated from patients with acute glomerulonephritis. J Exp Med 149:459–472

Villarreal H, Jr, Espinoza LR, Zabriskie JB (1982) Post-Bacterial glomerulonephritis. In: Zabriskie JE, Fillit H, Villarreal H, Becker EL (eds) Clinical immunology of the kidney. Wiley, New York, pp 239–262

Wilkinson AE (1984) Syphilis, rabbit syphilis, yans and pinta. In: Wilson G, Miles A, Parker HT (eds) Topley Wilson's principles of bacteriology, virology and immunity, vol 3. Williams and Wilkins, Baltimore, p 547

Winchester RJ, Kunkel HG (1979) The human Ia system: In: Kunkel HG, Dixon F (eds) Adv. immunology, vol 28. Academic, New York, pp 221–292

Yang LC, Soprey PR, Wittner MK, Fox EN (1977) Streptococcal induced cell mediated immune destruction of cardiac fibers in vitro. J Exp Med 146:344

Zabriskie JB (1982) Rheumatic fever: a streptococcal-induced autoimmune disease? Pediatr Ann 11:383–396

Zabriskie JB, Freimer EH (1966) An immunological relationship between the group A streptococcus and mammalian muscle. J Exp Med 124:661–678

Zabriskie JB, Hsu KC, Seegal BC (1970) Heart-reactive antibody associated with rheumatic fever: characterization and diagnostic significance. Clin Exp Immunol. 7:147–159

Genetic Approaches to the Study of Disease Resistance: With Special Emphasis on the Use of Recombinant Inbred Mice

D.E. Briles,[1] W.H. Benjamin, Jr.,[1] W.J. Huster,[2] and B. Posey[1]

1 Introduction

Central to studies of microbial pathogenesis is an identification of the protective mechanisms that defend the host against each pathogen. One way to do this is to identify genes of the host that affect the resistance to the pathogen and then determine their mechanisms of action. This approach automatically establishes the in vivo relevance of all mechanisms studied. the discovery of variant alleles affecting particular functions also serves to identify important protective mechanisms that might have otherwise remained unknown. The power of this type of approach is aptly illustrated by the major histocompatibility complex of mammals and birds, where much of our present knowledge has come from attempts to understand how a single genetic region could affect graft rejection, immunoresponsiveness, and disease susceptibility.

[1] Cellular Immunobiology Unit of the Tumor Institute, Department of Microbiology, and the Comprehensive Cancer Center, University of Alabama at Birmingham, Birmingham, Alabama 35294, USA
[2] Cellular Immunobiology Unit of the Tumor Institute, Department of Biomathematics and Biostatistics, and the Comprehensive Cancer Center, University of Alabama at Birmingham, Birmingham, Alabama 35294, USA

Current Topics in Microbiology and Immunology, Vol. 124
© Springer-Verlag Berlin·Heidelberg 1986

In an attempt to discover the essential mechanisms accounting for particular phenotypes, genes that have biological effects of interest should be identified and distinguished from each other whenever possible. The separate identity of each gene is established, in part, by mapping each gene under study into a genetic map of the organism's genome. Genes with different map positions are clearly different. Noncomplementing genes resulting in the same phenotype and occupying the same apparent map position are assumed to be alleles.

One of the most useful systems for determining the map position of new genes in the mouse is to determine the distribution of the phenotype under study in a panel of recombinant inbred mouse strains (BAILEY 1981). Recombinant inbred strains are a group of inbred lines derived from F2 progeny of a pair of parental inbred strains. Because they are inbred, each RI line is considered to be homozygous at each locus. Thus at any locus for which the original parental strains differed each RI strain will be homozygous for one or the other of the two parental alleles. Since the lines are developed in the absence of overt genetic selection, the particular homozygous allele at each locus is determined only by random segregation and crossover events of meiosis.

1.1 Recombinant Inbred Strains

If a set of RI strains (produced from a single pair of parental lines) can be divided into two distinct phenotypic groups when examined for a trait of interest, it is likely that the differences in phenotype among the RI strains are controlled by a single genetic locus. In an attempt to determine the approximate map position of the hypothetical locus, the strain distribution of the phenotypic trait is compared with the strain distribution of alleles at loci for which the RI strains have already been typed. If the phenotypic difference is found to be closely associated with the inheritance of a single chromosomal region, it provides both an approximate map position for a gene affecting the phenotype in question and strong evidence that the phenotypic difference being studied is affected by a single variant genetic locus in the set of RI strains being used.

BAILEY has written an excellent review of the use of RI strains, which lists many of the available strains and includes a description of the appropriate methodology and mathematical treatment of data involving single gene inheritance (BAILEY 1981).

1.2 Multi-Loci Inheritance

It has been known for some time that the overall resistance of an animal to any particular infectious agent is generally the result of the cumulative effect of genes at multiple loci. Examples of other traits controlled by multiple loci are height, weight, longevity, milk production, disease resistance (WRIGHT 1968; SNYDER and DAVID 1957), and immune responsiveness (BAKER et al. 1978; BRILES et al. 1977; NOWACK et al. 1975; MCCARTHY and DUTTON 1975; BIOZZI et al. 1975). In fact, for most traits where there are not discrete phenotypic groupings in the population, inheritance is usually found to be multigenic. Such traits have been referred to as "quantitative traits." Traits where individuals can be easily divided into phenotypic groupings are referred to as "qualitative" and generally are found to be controlled by one, or at most a few, loci. Examples of qualitative traits include albinism, blood groups, and enzyme polymorphisms.

In spite of the fact that immune responsiveness and resistance to diseases are quantitative traits, some of the genes affecting them have effects large enough to have allowed them to be easily detected in genetic backcross studies as well as with recombinant inbred mice. Such genes would include *nu* (PRITCHARD et al. 1973), *H-2* (SCHREFFLER and DAVID 1975), *Ig* (BLOMBERG et al. 1972; RI-HOVA-SKAROVA and RIHA 1974; McCARTHY and DUTTON 1975; BRILES et al. 1977), and *xid* (SCHER 1982), which affect immune responsiveness; and *xid* (O'BRIEN et al. 1979), *lps* (O'BRIEN et al. 1980a; SVANBORG EDEN et al. 1985), and *Ity* (PLANT et al. 1982), which affect the susceptibility of mice to certain pathogens.

2 Genetic Approaches for the Study of "Quantitative" Salmonella Resistance Genes

The genetics of genes with smaller effects on quantitative traits can be difficult to study using breeding studies or RI mice. However, what these genes tell us about protective mechanisms may be no less important than what is learned by genes with larger effects. The key to the analysis of the multiple genes involved in quantitative traits is to devise experimental approaches that allow the effects of some genes to be studied in the absence of the effects of others. General approaches for doing this are outlined in Sect. 2.1–2.4 below.

2.1 Examination of a Subset of the Disease Processes

One means of simplifying the genetics is to study only a portion of the disease process. In the case of studies of salmonella pathogenesis, this is the type of approach that is taken when mice are infected i.v. rather than orally and the data are collected by looking at the numbers of organisms in the liver or spleen at a specific time postinjection (HORMAECHE 1979) rather than determining the oral dose required to cause death. Using this approach, the influences of genes affecting the localization and subsequent net growth of salmonella in the liver or spleen can be detected in the absence of influences by genes affecting bacterial growth in the gut, bacterial entrance into the blood or lymph, and the "toxicity" of the salmonella for the host. Genes affecting these latter traits could be studied by designing experiments specifically to examine those portions of the disease process.

Portions of the disease process may be selectively studied by conducting studies in vitro, as has been done with salmonella-infected macrophages (LISSNER et al. 1983).

2.2 Use of Congenic Mice

Another strategy would be to compare the disease susceptibility of congenic mice. Such strains differ in only the small regions of DNA surrounding an individual "marker" locus. These strains are usually made by backcrossing a known allele from one strain for ten or more generations onto the genetic background of an inbred line carrying a different allele at the same locus.

Comparisons of the bacterial resistance of pairs of appropriate congenic mouse strains should allow the detection of the effects of individual loci affecting bacterial resistance in the absence of effects of bacterial resistance genes at other loci (POTTER et al. 1983). A search for congenic mouse strains carrying bacterial resistance genes could easily be undertaken with existing strains of mice that have been made congenic for a large number of different marker loci. Over 700 congenic strains, mostly on the C57BL backbround, are available through Jackson Laboratories, Bar Harbor, ME (Greenhouse 1984).

By chance, some of these congenic strains would undoubtedly be found also to carry variant bacterial resistance loci either as the marker locus, or more likely in the DNA flanking the marker locus. Strains congenic for bacterial resistance genes would be detected by comparing the susceptibility of each congenic strain with that of the parental strain onto whose genetic background the marker loci had been backcrossed.

There are several reasons why this approach would not be expected to identify all of the interesting genes: (1) all important disease resistance loci are probably not linked closely enough to marker genes to be included in the DNA carried by the existing strains of congenic mice; (2) even if a resistance locus is linked closely enough to a marker locus to be included in the donor DNA carried by a congenic strain, the locus would not be identified unless the donor and recipient mouse strains carried different resistance alleles at the locus; and (3) any gene whose expression requires the presence of nonrecipient strain alleles at other loci would not be detected.

Some of the most interesting congenic strains that could be used for this purpose are the bilineal congenic strains (BAILEY 1981). Bilineal congenic strains differ from normal congenic strains in that all of the congenic strains in a set are made by using the same donor and recipient mouse strains. Thus, if donor strain B and background strain A differ in disease resistance, and if enough congenic A.B strains exist, it should be possible to use the congenic strains to identify the genes responsible for the differences in disease resistance between strains A and B. Bailey has calculated that 80 or more bilineal congenic strains would be required to cover the entire mouse genome (BAILEY 1981). Two of the largest sets of bilineal congenic strains are those made by crossing DBA/2 genes into BALB/c (POTTER et al. 1983) and BALB/c genes into C57BL/6 (BAILEY 1981).

2.3 Use of Congenic Pathogens

One way of focusing a genetic study on particular bacterial resistance genes is by careful selection of the bacterial isolate used for infection. The validity of this approach is apparent when one examines studies where different bacterial strains of the same species are used to infect a panel of different inbred or recombinant inbred mouse strains. We have recently conducted several studies of this type using *S. typhimurium*, where we determined the net in vivo growth of several different salmonella strains in a panel of mouse strains (BENJAMIN et al. 1986). We observed that the relative bacterial resistance of the different mouse strains, when compared with each other, was dependent on which bacterial isolates are used to infect them. This could only have been the case if there were genetic differences among the mouse strains as well as among the bacterial isolates. Furthermore these data indicated that the overall pathogenesis of the salmonella in the mice was the result of specific interactions between particular mouse and bacterial genotypes.

In some recent experiments (BENJAMIN et al. 1986), we observed that with some strains of *S. typhimurium*, the Ity^s and Ity^r alleles failed to affect the numbers of salmonella in the livers and spleens 6 days postinfection, even though those same alleles affected the virulence of other salmonella isolates. The salmonella strains that were not affected by differences at the Ity locus still showed large differences in net growth among different inbred mouse strains. These salmonella strains should be ideal for detecting non-Ity salmonella resistance genes of mice since they would avoid any effects due to genetic segregation at the Ity locus.

Particularly useful bacterial strains can be generated by making isogenic bacteria. This can be done by using conjugation or transduction to transfer specific virulence genes from one bacterial strain to another. When the relative pathogenesis of the isogenic bacterial strains are compared in a panel of inbred mice, the only differences in their relative pathogenicity in different inbred strains should be due to differences in the resistance genes of the mice that affect the virulence trait for which the bacteria are isogenic.

The validity of this approach appears to be confirmed by a recent study where we have used two salmonella strains congenic for a gene (or genes) required to exploit the Ity^s phenotype of mice (BENJAMIN et al. 1986) (see also Sect. 2.4.2).

The biochemical function of the virulence genes that exploit Ity^s is not yet known. There exist, however, a large number of genes in *Salmonella* and other bacteria whose biochemical functions are known. Some of these may prove useful in studies of mouse salmonella resistance genes. One important group of useful salmonella strains are those which carry mutant genes, such as *aroA* (HOISETH and STOCKER 1981; see also Chap. by Stocker), in metabolic pathways that are essential for in vivo growth. Mutants of this type should not be affected by mouse genes regulating bacterial growth. Thus enumeration of the number of colony-forming units (CFUs) of such a mutant in the liver or spleen of mice should be an ideal means of examining mouse genes that control bacterial killing.

2.4 Use of Recombinant Inbred Mice

One approach of looking for linkage of genes coding for disease resistance is to classify all of the infected recombinant inbred strains as being either "resis-

Table 1. Association of disease resistance with known genetic markers in RI strains

---

A. Calculate mean resistant/susceptibility parameter for each strain, after an appropriate data transformation

    1. Geometric mean CFU/spleen
    2. Reciprocal mean days to death
    3. Or etc.

B. Assignment of each RI strain to one of the two parental groups depending on its allele at the locus in question

C. Use of the Two-Sample Rank test to compare the mean resistance parameters of the individual RI strains in the two parental groups

D. Confirmation of suspected associations by use of classical genetics and/or congenic mice

---

tant" or "susceptible" based on the relative severity of disease. One then looks for an association between the segregation of known alleles in the strains and their newly assigned resistance and susceptible phenotypes.

This approach has worked well in situations where most of the variation in disease resistance among the various strains has been due to an allelic difference at a single locus. However, in most studies where recombinant inbred (RI) strains are used to examine disease resistance, the strains fail to form discrete "resistant" and "susceptible" groupings. In many studies the RI strains form a virtual continuum. When data of this type are encountered the standard techniques of using RI strains to look for linkage of qualitative traits are not appropriate because the information regarding the relative resistance of the strains within the resistant and susceptible populations is unavailable to the subsequent analysis.

To increase the amount of information from quantitative disease resistance data, we have developed a mathematical approach that allows us to calculate indexes of the associations of the alleles at each locus with disease resistance. Our approach takes into account the relative disease resistance of each RI strain to the disease resistance of each of the other strains (Table 1).

Our approach is based on the assumption that even in the presence of other segregating genes the RI strains inheriting any particular resistance allele will on average be more resistant than strains inheriting the allele for susceptibility at the same locus. A perfect correspondence between resistance and any particular genetic marker is not expected because of the effects of resistance genes segregating at other loci and because of crossovers between the marker gene and the bacterial resistance gene linked to it.

2.4.1 Transformation of Data Within each Strain

Our first step is to determine for each RI and parental strain a numerical score reflecting its relative disease resistance. This score is calculated in a manner appropriate to the distribution of the data. For example, if the data are obtained as the numbers of salmonella recovered 1 week postinfection, the geometric mean of the CFUs of salmonella recovered from the individual mice is used. If the data were collected as days to death postinoculation, we determine the reciprocal mean days to death (see Colwell et al., this volume). The reciprocal mean provides a mathematical approximation of median survival that is based on all of the data points. The reciprocal mean has an advantage over a standard medial survival which is based on only the central one or two values in each distribution.

2.4.2 Associations Between Loci and Resistance

Next, using the means of the individual strains as a measure of their resistance to infection, we use the two-sample rank test [also known as the two-sample rank test or the Mann-Whitney U-test (GOLDSTEIN 1964)] to compare the two groups of RI strains expressing the alternative parental alleles at each locus for which the distribution of parental alleles is known. By comparing the rank values obtained for each of the different loci, it is possible to identify those loci most likely to be linked to genes segregating for disease resistance. In this type of calculation the rank value provides a relative measure of association and should not be used to calculate literal probability values (see Sect. 2.4.3). Our reasons for choosing the two-sample rank test are given in Sect. 3.1 of this chapter.

As an illustration of this type of approach, we have analyzed some of the published data of O'Brien et al. where 26 B × D RI strains were infected subcutaneously with the TML strain of *S. typhimurium*. Figure 1 depicts the geometric mean number of salmonella recovered 9 days postinoculation from the three to seven mice tested from each strain, as reported by O'BRIEN et al. (1980b). We have plotted the RI and parental strains in order of increasing resistance. As can be seen, the strains appear to form a continuum in terms of the numbers of salmonella recovered from their spleens. This finding is common in cases of quantitative inheritance and is indicative of inheritance at more than one locus. When O'Brien and her colleagues analyzed their data, they felt that at least three loci, including *Ity*, probably accounted for the distributions of salmonella resistance seen among the RI strains of this panel.

2.4.2.1 Minimum Number of Loci Involved

The minimum number of loci required to cause any observed distribution of RI strain phenotypes can be calculated for quantitative traits using an equation derived by TAYLOR (1976): $L = D^2/2V$, where L is the minimal number of loci required, D is the difference between the most extreme RI strain values, and V is the variance of all RI strain values (see Sect. 3.2 of this chapter for more information about this method). This equation provides an unbiased estimate of the minimal number of loci required to cause any observed distribution of RI strain phenotypes. By applying Taylor's equation to the data in Table 2, an estimate of at least five different loci is obtained. This is in good agreement with the estimate of at least three loci made by O'BRIEN et al. (1980b).

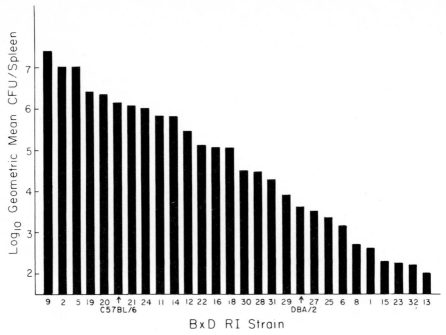

Fig. 1. The log mean TML CFU in spleens from 26 B × D recombinant inbred (RI) strains and their C57BL/6J and DBA/2 parental strains. Mice were killed 9 days postsubcutaneous inoculation. Data are taken from O'Brien et al. (1980b)

2.4.2.2 Sample Calculations

In an effort to identify some of these five or more loci, we have subjected O'Brien et al. (1980b) data to the rank analysis described above. The B × D RI strains have been typed for parental alleles at 107 loci which have been mapped to 12 of the 20 mouse chromosomes (Taylor, personal communication). Using our computer analysis we have been able to look for linkage of salmonella resistance genes to all of these 107 loci. For any particular comparison, a rank value of >1.96 has had a probability of occurrence by change of <0.05. For probabilities of $P<0.01$, $P<0.001$, and $P<0.0001$ the rank values are 2.8, 3.3, and 3.9 respectively. However, because of the large numbers of comparisons made statistically significant, rank values should not be taken as "proof" that a particular locus is associated with disease resistance but merely an indication of which loci should be subjected to further study.

When the RI mice were infected with TML, 8 of the 107 loci showed rank values of >1.96. Since six of these eight loci comprised closely linked pairs of genes in three different linkage groups, the inheritance of only five chromosomal regions appeared to be associated with the salmonella resistance of the B × D RI strains. These five loci are listed in Table 2 along with several other loci that have shown associations with the resistance of other salmonella strains. The *Igh-C* and *H-2* loci have been included as examples of the 98 other loci whose inheritance showed no associations with salmonella resistance.

Table 2. Test for associations between resistance to infection with TML *S. typhimurium* and individual loci segregating in B × D RI mice

Locus	Chromo- some	RI strains with allele from					
		C57BL		DBA/2		Rank values	
		No. RI strains	Log_{10} mean[a] CFU/spleen	No. RI strains	Log_{10} mean CFU/spleen	Observed data	Data normalized to *Ity*
Ity	1	15	5.77	11	3.00	4.1	0.0
Ly-9	1	13	4.34	11	5.16	1.2	2.8
cdm	3	14	4.32	12	4.93	0.8	0.2
Akp-2	4	13	5.32	13	3.88	2.3	1.5
Pgm-1	5	11	3.66	15	5.30	2.5	1.4
Ric	5	13	3.90	12	5.56	2.4	0.4
"B"	7	15	4.65	9	4.82	0.1	0.8
D12-1	12	8	5.47	16	3.97	2.1	1.5
Igh-C	12	12	4.51	14	4.68	0.1	0.4
Mtv-9	12	15	5.18	11	3.81	2.1	0.8
H-2	17	9	4.29	17	4.76	0.5	0.8

[a] Calculated from data of O'BRIEN et al. (1980)

In Fig. 2, we have plotted the means of the RI strains in each of the two parental groups for four of the loci listed in Table 2. Associations between the patterns of inheritance of the DBA/2 *Ity* locus and the C57BL *Pgm-1* locus and salmonella resistance are apparent. The existence of the *Ity* locus and its effect on salmonella resistance has been well established (PLANT et al. 1982). The fact that *Ity* inheritance does not correspond perfectly with resistance is thought to be due to effects of resistance genes at other loci (O'BRIEN et al. 1980b).

When one of the loci affecting a particular trait has as strong an effect as the *Ity* alleles, two types of problems can occur when looking for associations with other loci. One problem is that any other locus, whose alleles by chance show the same strain distribution as the strong locus, will erroneously appear to affect the trait in question. The other problem is that because of the effects of the strong locus, it will tend to mask any association between the trait being studied and genes with smaller effects. Both of these problems can be partially overcome by normalizing the data for any known resistance locus when examining the results at the other loci. In the case of the *Ity* locus, we have done this mathematically by multiplying all CFU values for the *Ity*^{s} strains by the factor that is required to reduce the rank value of the comparison between *Ity*^{s} and *Ity*^{r} strains to zero. The value of this correction factor is determined by a series of approximations using a computer program.

In the case of the RI mice infected with TML, this normalization for *Ity* eliminated the association between all four non-*Ity* loci and salmonella resistance, indicating that the weak associations seen in the nonnormalized data were spurious (Table 2). Quite surprisingly the normalization also visualized

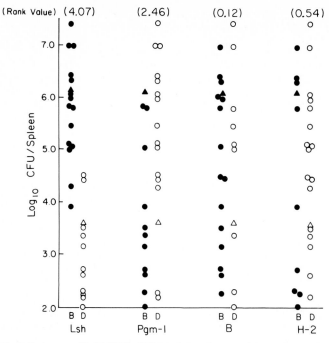

Fig. 2. Log mean SR-II CFU of RI (*circles*) and parental (*triangles*) strains of mice plotted according to their genotype at four different loci. *Solid symbols* indicate that the strain carries the allele from the C57BL/6 strain at the indicated locus. *Open circles* indicate that the strain carries the allele from the DBA/2 parent. Rank values of greater than 1.96 reflect a statistical difference in the distribution of strains carrying different parental alleles. Values for the parental strains C57BL/6 and DBA/2 are not used in the calculations of the rank values

a strong association between salmonella resistance and *Ly-9*, a gene at the distal end of chromosome 1. Although these data suggest that there may be a salmonella resistance gene at the distal end of chromosome 1, this has not been independently confirmed by use of congenic mice or by backcross analysis, and should be regarded as tentative.

For the purposes of illustration, we have analyzed another set of data that was obtained when mice were infected with salmonella strain WB500. This strain is a transconjugant from a mating between SR-11 and LT2-Z. In this set of data, only 17 of the B × D strains were used. As in the case of TML, the resistance to infection with this salmonella strain showed an association with *Ity*. In addition, however, an association was also seen with the loci *cdm* and "B". When the data for infection with WB500 were normalized for the effect of *Ity*, the effect of the *cdm* locus actually increased. This indicated that the effect of the *cdm* locus was not because of a chance cosegregation of the *cdm* and *Ity* genes in the RI strains, and may thus be due to some gene linked to *cdm*. As in the case of the association with *Ly-9*, the association with disease resistance and *cdm* should be regarded as tentative.

In addition to examining the relationship between CFUs from the different RI strains and the inheritance of different marker loci, a very powerful approach

Table 3. Test for associations between resistance to infection with *S. typhimurium* and individual loci segregating in B × D RI mice. Rank values based on

CFU WB500			CFU WB500/CFU LT2-Z	
Observed data		Normalized to *Ity*	Observed data	Normalized *Ity*
2.1	*Ity*	0.0	3.4	0.1
1.3	*Ly-9*	0.7	0.4	0.1
2.3	*cdm*	2.7	0.7	1.8
0.8	*Akp-2*	0.9	1.3	0.9
0.4	*Pgm-1*	1.4	0.6	0.2
1.7	*Ric*	0.4	2.8	0.4
2.5	"B"	1.9	0.7	0.3
1.15	*D12-1*	0.7	1.3	0.6
0.5	*Igh-C*	0.8	1.2	1.1
0.5	*Mtv-9*	0.1	0.4	1.3
0.1	*H-2*	0.6	0.8	0.4

can be to examine the association between the marker loci and the relative ability of two different bacterial strains to cause infection. In this section we describe a comparison of the relative growth rates of LT2-Z and WB500 in the B × D RI mice. Since the only difference between LT2-Z and WB500 are the genes LT2-Z got from SR-11, any difference in the growth rates of LT2-Z and WB500 in any particular strain should be due to the genes LT2-Z received from SR-11. Furthermore any differences between the relative growth of WB500 and LT2-Z in different RI strains should be due to interactions between the genes LT2-Z inherited from SR-11 and the salmonella resistance genes that had segregated among the RI strains.

We had previously shown that the growth of LT2-Z in B × D mice was not regulated by the alleles of the *Ity* locus (rank value of only 1.15). In preliminary studies, using standard inbred strains, it appeared that WB500 contained an SR-11 gene, or genes, that could allow salmonella with LT2-Z genotype to exploit the salmonella-susceptible phenotype of Ity^s mice. Thus we expected the relative net growth of WB500 and LT2-Z salmonella in B × D mice to show an association with the *Ity* alleles of these mice.

In Table 3, it can be seen that the relative growth of WB500 versus LT2-Z (expressed as CFU WB500/CFU LT2-Z) showed an association with *Ric* and an even stronger association with *Ity* than did the net growth of WB500 (Table 2). When these data were normalized for *Ity*, the association with *Ric* vanished and no new associations appeared at other loci. This was consistent with our hypothesis that the extra virulence genes carried by WB500 act specifically to exploit the *Ity* locus.

2.4.3 Limitations of the Rank Order Approach

A major caution must be raised about the interpretation of the data presented in Tables 2 and 3. The rank test should only be used to indicate which loci have the highest probability of being linked to disease resistance loci. It cannot be concluded that any particular locus is really associated with salmonella resistance without confirmation by backcross analysis or testing with appropriate congenic mice. In most situations where the two-sample rank test is used, the rank values can be used to look up P values in a table giving the area under a normal curve.

Any P values obtained from such a table would not be valid for the type of comparisons made here. The reason for this is that because of the large number of comparisons (one for each locus examined) some significant P values would be expected by chance alone. For example, out of 100 independent comparisons, five would expect to occur at the $P < 0.05$ level by chance alone. Determining the expected number of significant P values is not a straightforward calculation since many of the comparisons in this type of analysis are not independent. This is because many of the loci are linked closely enough that they cosegregated with at least one other locus in most or all of the RI strains. (In the present data set the 107 loci examined fall into 47 closely linked groups of genes.) This would act to reduce the number of "significant" associations occurring by chance above. Thus the effective number of independent comparisons in our study should be less than the 107 actually made.

Although the rank approach can be applied to all sets of RI strains, the value of this type of approach is significantly enhanced as the number of RI strains used increases above 20–30 strains.

2.4.4 Computer Program

The calculations shown in Tables 2 and 3 have all been done with the programs we have written for use with an Apple IIe computer. Since the main program occupies more than 150 lines it has not been reproduced here. We will, however, gladly send a listing of the program to anyone requesting one, or a program copy to any one sending us a blank 5 1/2" diskette.

3 Additional Statistical Considerations

3.1 Advantages of the Two-Sample Rank Test

The Two-Sample Rank test has been used since it makes a nonparametric analysis of the data. This is superior for this purpose to a parametric test, such as the t-test or the analysis of variance, since with parametric tests any deviation of the values within each group of strains from a normal distribution would lead to an underestimate of the statistical significance of any differences between the two groups of strains. Parametric tests are only appropriate for data where the values are selected from a single population with a normal distribution. A normal distribution has easily discernable properties; the data points fall symmetrically around the mean such that the mean is the most frequently observed value and is the median of all values.

In some cases, it is possible to transform the data to a normal distribution by using the appropriate transformation. This is apparently the case for the data in Tables 2 and 3, since, if a log transformation of those data is performed the P values calculated from a Student's t-test are virtually identical to those calculated with the rank test. In other situations this is not possible. Data are unlikely to form a normal distribution when they consist of the number of days alive postinoculation. The average days alive for the RI strains surviving infection and the RI strains that die of the infection would most likely form separate modes of a bimodal distribution. Data in a bimodal distribution cannot be transformed to a normal distribution by any procedure.

In our computer program, we have chosen to use the rank value as our index of association between resistance and the marker loci. By using the rank test, we can be confident that our calculations of association between marker genes and disease resistance will be optimal and will not be dependent on whether the data can be transformed to a normal distribution or whether an appropriate transformation of the data is used in the calculations.

Standard procedures for multivariant analysis have not been used because the large number of possible genotypes prohibits them from giving meaningful results. For example, in a hierarchical nested analysis of variance, the number of different cells at the bottom of the hierarchy would be 2^n, where n is the number of independently segregating loci. If only seven loci were examined, the number of cells could be 128. To ensure a valid nested analysis of variance, one would need several RI strains in each cell or a total of at least several hundred RI strains, many more than the 10–30 usually available.

3.2 Estimation of the Minimum Number of Loci by Segregation Analysis: Taylor's Formula

This equation (Sect. 2.4.2.1) provides an estimate of the number of genes involved in a multigenic trait segregating among a set of RI strains. It is based on several assumptions (TAYLOR 1976; BAILEY 1981):

1. Out of all possible genotypes, the two that express the highest and lowest mean values are included among the strains studied.
2. Of the highest and lowest strains, both or neither may be progenitor strains, but not just one.
3. The loci have equal effects on the trait.
4. The loci are unlinked.
5. There is no epistacy.

It is very unlikely that more than one or two of these assumptions would be correct in any particular experiment. In such a case, the calculated number of loci would be less than the real number. Thus this calculation provides only a minimal estimate of the number of loci involved in a quantitative trait. A more accurate estimate of the number of genes involved in a particular trait can be obtained if enough mice of each strain are infected to determine the number of phenotypic groups into which the strains fall. When the number of phenotypic groups is known, the equation that should be used is $L = \log (F + 2)/\log 2$ (BAILEY 1981), where $L =$ the number of loci and $F =$ the number of phenotypic groups. Although this is a more precise measure of the number of genes, it would require that enough mice be tested of each RI strain to enable each phenotypic group to be statistically distinguished from every other phenotypic group. Because of the large numbers of mice required for this type of experiment, the costs incurred probably would not be justified.

Acknowledgments. This research has been supported by grants CA 16673 and CA 13148 awarded by the National Cancer Institute and AI 15986 awarded by the National Institute of Allergy and Infectious Diseases and grant 420 awarded by the Alabama Research Institute. David E. Briles is the recipient of a Research Career Development Award, AI 00498.

We would like to acknowledge the advice of Dr. S.-J. Soony of the Department of Biomathematics and Biostatistics at the University of Alabama at Birmingham. We are also grateful for the assistance of Ann Brookshire and Carol McNeeley for preparation of the manuscript, Maxine Aycock for preparation of the illustrations, and Carolyn Booker for her critical review of the manuscript.

References

Bailey DW (1981) Recombinant inbred strains and bilineal congenic strains. In: Fuster HL, Small JD, Fox JG (eds) The mouse in biomedical research, vol 1. Academic, New York, pp 224–239

Baker PJ, Amsbaugh DF, Prescott B, Stashak PW, Rudbach JA (1978) Multigenic control of the antibody response to type III pneumococcal polysaccharide and other helper T cell independent antigens. In: Friedman H, Linna TJ, Prier JE (eds) Infection immunity and genetics. University Park Press, Baltimore, pp 67–84

Benjamin W, Turnbough C, Posey B, Briles DE (1986) *Salmonella typhimurium* virulence genes necessary to exploit the $Ity^{s/s}$ genotype of the mouse. Infect Immun 51(3)

Biozzi G, Stiffel C, Mouton D, Bouthillier Y (1975) Selection of lines of mice with high and low antibody responses to complex immunogens. In: Benacerraf B (ed) Immunogenetics and immuno-deficiency. University Park Press, Baltimore, pp 180–227

Blomberg B, Geckeler W, Weigert M (1972) Genetics of the antibody response to dextran in mice. Science 177:178–180

Briles DE, Krause RM, Davie JM (1977) Immune response deficiency of BSVS mice. I. Identification of Ir gene differences between A/J and BSVS mice in the anti-streptococcal group A carbohydrate response. Immunogenetics 4:381–392

Goldstein A (1964) Biostatistics: an introductory text. MacMillan, New York, pp 55–59

Greenhouse DD (1984) Holders of inbred and mutant mice in the United States. National Academy, Washington

Hoiseth SK, Stocker BAD (1981) Aromatic-dependent *Salmonella typhimurium* are non-virulent and effective as live vaccines. Nature (London) 291:238–239

Hormaeche CE (1979) Natural resistance to *Salmonella typhimurium* in different inbred mouse strains. Immunology 37:311–318

Lissner CR, Swanson RN, O'Brien AD (1983) Genetic control of the innate resistance of mice to *Salmonella typhimurium*: expression of the *Ity* gene in peritoneal and splenic macrophages in vitro. J Immunol 131:3006–3013

McCarthy MM, Dutton RW (1975) The humoral response of mouse spleen cells to two types of sheep erythrocytes. I. Genetic control of the response to H and L SRBC. J Immunol 115:1316–1321

Nowack H, Hahn E, David CS, Timpl R, Gotze D (1975) Immune response to calf collagen type I in mice: a combined control of Ir-1A and non-H-2 linked genes. Immunogenetics 2:331–335

O'Brien AD, Scher I, Campbell GH, MacDermott RP, Formal SB (1979) Susceptibility of CBA/N mice to infection with *Salmonella typhimurium*: influence of the X-linked gene controlling B lymphocytes function. J Immunol 123:720–724

O'Brien AD, Rosenstreich DL, Scher I, Campbell GH, MacDermott RP, Formal SB (1980a) Genetic control of susceptibility to *Salmonella typhimurium* infection in mice: role of the *lps* gene. J Immunol 124:20–24

O'Brien AD, Rosenstreich DL, Taylor BA (1980b) Control of natural resistance to *Salmonella typhimurium* and *Leishmania donovani* in mice by closely linked but distinct genetic loci. Nature 287:440–442

Plant JE, Blackwell JM, O'Brien AD, Bradley DJ, Glynn AA (1982) Are the *Lsh* and *Ity* disease resistance genes at one locus on mouse chromosome 1. Nature 297:510–511

Potter M, O'Brien AD, Skamene E, Gros P, Forget A, Kongshaun PAL, Wax JS (1983) A BALB/c congenic strain of mice that carries a genetic locus (*Ity*^{r}) controlling resistance to intracellular parasites. Infect Immun 40:1234–1235

Pritchard H, Riddaway J, Micklem HS (1973) Immune responses in congenitally thymus-less mice. Clin Exp Immunol 13:125–138

Rihova-Skarova B, Riha I (1974) Genetic regulation of the immune response to haptens. Ann Immunol (Paris) 125C:195–198

Scher I (1982) The CBA/N mouse strain: an experimental model illustrating the influence of the X-chromosome on immunity. Adv Immunol 33:1–71

Schreffler DC, David CS (1975) The major histocompatibility complex and the I immune response region. Adv Immunol 20:125–195

Snyder LH, David PR (1957) The principles of heredity. Heath, Boston, pp 197–210

Svanborg Eden C, Briles DE, Hagberg L, Michalek SM, McGhee JR (1985) Susceptibility to *Escherichia coli* urinary tract infection linked to the *Lps*^{d} gene. J Immunol (to be published)

Taylor BA (1976) Genetic analysis of susceptibility to isoniazid-induced seizures in mice. Genetics 83:373–377

Wright S (1968) Evolution and genetics of populations, vol I. Genetic and biometric foundations. University of Chicago Press, Chicago, pp 372–420

Influence of Host Genes on Resistance of Inbred Mice to Lethal Infection with *Salmonella typhimurium*

A.D. O'BRIEN

1 Introduction

Salmonella typhi, the causative agent of typhoid fever, is avirulent for mice. The parenteral 50% lethal dose (LD_{50}) of *S. typhi* for mice of all inbred strains examined to date is $\geq 10^8$ bacteria (GERICHTER 1960; CARTER and COLLINS 1974a; O'BRIEN 1982) unless the animals are pretreated with iron and/or an iron chelator (POWELL et al. 1980; O'BRIEN 1982) or the mice are challenged intraperitoneally with *S. typhi* suspended in hog gastric mucin (NUNGESTER et al. 1936; SPAUN 1964). By contrast, *Salmonella typhimurium* primarily evokes gastroenteritis in man, but causes a typhoid fever-like disease in mice (called murine typhoid). Both human typhoid and murine typhoid are systemic illnesses. In mice infected orally with *S. typhimurium* or in humans who ingest food or water contaminated with *S. typhi*, the bacteria either multiply in the small bowel or directly penetrate the intestinal mucosa without apparent enteric colonization (GERICHTER 1960; HORNICK et al. 1970; HOHMAN 1978). Studies in mice have established that the foci from which salmonellae disseminate are the Peyer's patches of the small intestine (CARTER and COLLINS 1974b; HOHMAN et al. 1978). The bacteria apparently gain access to the circulation via the lymphatics, seed the reticuloendothelial cell system (RES), and replicate within splenic and hepatic tissues. A secondary bacteremia ensues following growth of salmonellae in these RES organs which, in turn, furthers systemic dissemina-

Department of Microbiology, Uniformed Services University of the Health Sciences, 4301 Jones Bridge Road, Bethesda, Maryland 20814, USA

Current Topics in Microbiology and Immunology, Vol. 124

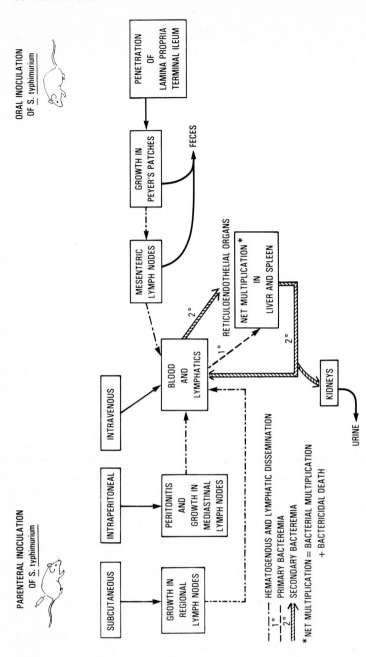

Fig. 1. The pathogenesis of murine typhoid

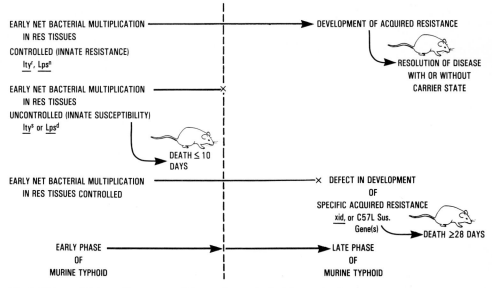

Fig. 2. Effects of *Salmonella* response alleles on the survival of the murine host

tion of the organisms. The pathogenesis of murine typhoid is illustrated in Fig. 1. It should be emphasized that the sequence of events that occurs once *S. typhimurium* becomes systemic is the same whether animals are inoculated orally or parenterally. Indeed, STUART (1970) states the disease (salmonellosis) in mice is a septicemia and is of no importance as an infection of the gastrointestinal tract. In all experimental work it is important to control dosage as accurately as possible and i.p. or i.v. injection of the infective dose eliminates variables and gives rise to a disease having all the important characteristics of the natural infection.

Several experimental parameters influence whether mice survive challenge with *S. typhimurium*, and these include: the strain of *S. typhimurium*, the dose and route of infection, and the particular strain of mice (HORMAECHE 1979a; O'BRIEN et al. 1983; PLANT 1983). When mice of different inbred strains are inoculated via the same route, they exhibit dose-dependent variable susceptibility to *S. typhimurium* (ROBSON and VAS 1972; PLANT and GLYNN 1976; HORMAECHE 1979a); the parenteral LD_{50} of highly virulent *S. typhimurium* for some strains (e.g., BALB/c, C57BL/6) is ≤ 10 bacteria, whereas the LD_{50} for other strains (e.g., CBA, A/J) is $\geq 10^4$ (reviewed by O'BRIEN et al. 1980a; ROSENSTREICH et al. 1982). The differential response of mice to *S. typhimurium* may therefore serve as a model for the genetic control of resistance to an infectious organism and as a probe to evaluate mechanisms of immunity to typhoid fever. Furthermore, recent studies indicate that expression of several distinct host genes determines whether *S. typhimurium*-inoculated mice survive murine typhoid (O'BRIEN et al. 1983), and the genes appear to act at different phases of the infectious process. Thus, a careful dissection of how specific host genes affect the response of mice to *S. typhimurium* infection will facilitate a better understanding of the importance of immunologically specific and nonspecific resistance mechanisms during each stage of murine typhoid. A division of murine salmonellosis into different phases and the kinds of host defenses required to control the growth of *S. typhimurium* at these various times during the progression of the disease are depicted in Fig. 2.

2 Early-Stage Salmonella Response Genes

2.1 *Ity*: Genetics

The host gene which regulates how well mice control the early replication of *S. typhimurium* in splenic and hepatic tissues after intravenous (i.v.) or subcutaneous (s.c.) challenge with *S. typhimurium* was designated *Ity* (for immunity to *typhimurium*) by PLANT and GLYNN (1977). These investigators recognized the existence of *Ity* by the response to s.c. challenge of F_1, F_2, and backcross generations derived from matings of salmonella-resistant (LD_{50} of *S. typhimurium* C5 by s.c. route for CBA mice $= 1 \times 10^7$) and salmonella-susceptible (LD_{50} of *S. typhimurium* C5 by s.c. route for BALB/c mice < 10) parental strains (1976). They observed that resistance among these hybrid mice was controlled by a single, autosomal, non-H-2-linked gene, and they called the susceptibility allele Ity^s and the resistance allele Ity^r. PLANT and GLYNN subsequently mapped *Ity* to mouse chromosome 1 by a series of linkage studies with known chromosomal markers (PLANT and GLYNN 1979). We confirmed that the *Ity* locus was located on mouse chromosome 1 by an evaluation of the pattern of salmonella susceptibility among 48 recombinant inbred strains of mice (O'BRIEN et al. 1980b). Our data also indicated that *Ity* was closely linked but distinct from the gene designated *Lsh* that controls the extent to which *Leishmania donovani* replicates in the RES of mice during the first few weeks of parasitic disease. Our rationale for considering *Ity* and *Lsh* discrete loci was twofold: three recombinant inbred mouse strains exhibited discordant responses to *S. typhimurium* and *L. donovani*; and the mapping studies of PLANT and GLYNN (1979) and BRADLEY et al. (1979) had placed the two loci, *Ity* and *Lsh*, on opposite sides of the *Idh-1* (isocitrate dehydrogenase) chromosome 1 marker. However, a re-examination of the three recombinant inbred strains with putatively different *Lsh* and *Ity* phenotypes and of various backcross generations derived from them showed no discordance between *Lsh* and *Ity* typings (PLANT et al. 1982). Subsequently, we used a five-point backcross to establish the precise location of *Ity/Lsh* as between *Idh-1* and the color-coat marker leaden (TAYLOR and O'BRIEN 1982). When taken together, these new observations indicate that *Ity* and *Lsh* are either so closely linked as to be indistinguishable or are in fact identical loci. A report by SKAMENE et al. (1982) favored the latter interpretation. These researchers examined the responses of 14 inbred and 38 recombinant inbred strains of mice to *Mycobacterium bovis* bacille Calmette-Guérin (BCG) infection and categorized mice as Bcg^s or Bcg^r according to whether the microbe replicated rapidly in RES tissues (*Bcg*^{s} phenotype) or whether growth was restricted (*Bcg*^{r} phenotype). SKAMENE et al. (1982) found that the distribution of *Bcg*^{s} and *Bcg*^{r} alleles exactly matched the patterns of *S. typhimurium* and *L. donovani* responses and suggested that resistance to all three microorganisms is regulated by expression of a single gene. Murine resistance to *M. lepraemurium* may also be controlled by *Ity* (BROWN et al. 1982).

2.2 *Ity*: Expression

We recently monitored the kinetics of *Ity* expression in *Ity*^{r} and *Ity*^{s} mice challenged intravenously with *S. typhimurium* (SWANSON and O'BRIEN 1983). The purpose of our study was to assess how soon after bacterial inoculation differences in net growth of salmonellae in the RES of *Ity*^{s} and *Ity*^{r} could be detected and to evaluate the in vivo interaction between *S. typhimurium* and RES macrophages. At the time this study was initiated, we and other investigators (HORMAECHE 1979a; 1979b; O'BRIEN et al. 1979a; PLANT et al. 1982) believed on the basis of circumstantial evidence that the effector cell for *Ity* expression was the resident murine macrophage. The findings in support of this theory were as follows: (1) the host cell(s) that confers the *Ity*^{r} phenotype is sensitive to silica (O'BRIEN et al. 1979a), an agent toxic to macrophages (KESSEL et al. 1963); (2) mature T cells are not required for restriction of early *S. typhimurium* growth by *Ity*^{r} mice (METCALF and O'BRIEN and METCALF 1982); (3) MAIER and ÖELS (1972) reported that peritoneal macrophages from BRVR mice, animals which are now believed to be *Ity*^{r} (BENJAMIN and BRILES 1981; O'BRIEN et al. 1982b), kill *S. typhimurium* better than do macrophages from the salmonella-susceptible BSVS strain of mice. BLUMENSTOCK and JANN (1981) in a later report reached the same conclusion in experiments with salmonella-infected peritoneal macrophages from *Ity*^{r} (C3H/Hf) and *Ity*^{s} (C57BL/6J) mice.

In our in vivo studies with *Ity*^{s} and *Ity*^{r} (SWANSON and O'BRIEN 1983), 99% of the *S. typhimurium* were cleared from the blood of both kinds of mice within 2 h of intravenous injection, and uptake of the bacteria by splenic and hepatic tissue (presumably by the resident macrophages in these organs) was similar regardless of *Ity* genotype. The fate of *S. typhimurium* in the RES of these mice then followed a biphasic course during the first 24 h after challenge. During the first phase (0–4 h), only killing of salmonellae was evident, and this microbicidal activity was independent of *Ity* type. In the second phase (4–24 h), the number of *S. typhimurium* in the livers and spleens of both *Ity*^{s} and *Ity*^{r} mice increased, but the bacteria accumulated faster in *Ity*^{s} than in *Ity*^{r} RES organs. That this more rapid net growth of *S. typhimurium* in inbred mice of the *Ity*^{s} versus *Ity*^{r} phenotype was in fact due to expression of the *Ity* gene was confirmed by a comparison of the in vivo net multiplication rate of salmonellae in BALB/c (*Ity*^{s}) and mice congenic except for *Ity*^{s}. These congenic mice, bred by POTTER et al. (1983) and designated C.D2 *Ity*^{r}, are BALB/c mice with the DBA/2 *Ity*^{r} allele and a 30 centimorgan segment of DBA chromosome that surrounds that locus.

To demonstrate directly that the effector cell for *Ity* expression is the resident murine macrophage and to examine further the phenotypic expression of this

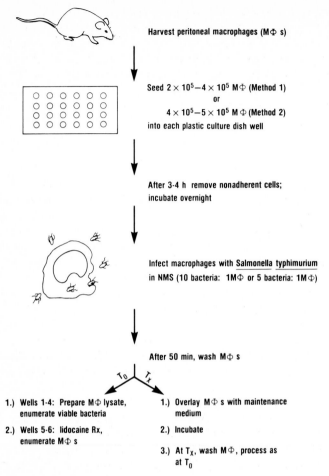

Harvest peritoneal macrophages (MΦ s)

Seed $2 \times 10^5 - 4 \times 10^5$ MΦ (Method 1)
or
$4 \times 10^5 - 5 \times 10^5$ MΦ (Method 2)
into each plastic culture dish well

After 3-4 h remove nonadherent cells;
incubate overnight

Infect macrophages with <u>Salmonella typhimurium</u>
in NMS (10 bacteria: 1MΦ or 5 bacteria: 1MΦ)

After 50 min, wash MΦ s

T_0 T_X

1.) Wells 1-4: Prepare MΦ lysate,
    enumerate viable bacteria

2.) Wells 5-6: lidocaine Rx,
    enumerate MΦ s

1.) Overlay MΦ s with maintenance
    medium

2.) Incubate

3.) At T_X, wash MΦ, process as
    at T_0

Fig. 3. In vitro macrophage assay

gene, we developed an assay to compare the kinetics of *S. typhimurium* growth within *Ity*^r and *Ity*^s macrophages isolated in vitro (LISSNER et al. 1983). Details of the assay are summarized in Fig. 3. With our in vitro macrophage assay we made the following observations. First, the *Ity* phenotype of genetically inbred strains of mice was expressed by both *S. typhimurium*-infected resident peritoneal macrophages and resident splenic macrophages. Second, the more extensive net growth of *S. typhimurium* in the RES of *Ity*^s mice by 24 h after infection was found to correlate with differential net growth of salmonellae in resident peritoneal macrophages from mice congenic (BALB/c versus C.D2 *Ity*^r mice) except at the *Ity* locus. Furthermore, no difference in phagocytosis of radiolabeled *S. typhimurium* by macrophages from BALB/c and C.D2 *Ity*^r mice was noted. Third, macrophages from C.D2 *Ity*^r mice infected with a temperature-sensitive mutant of *S. typhimurium* killed these nonreplicating salmonellae more efficiently than did BALB/c macrophages. Hence, we conclusively demonstrated that the effector cell for *Ity* expression is the resident macrophage, and our findings strongly indicated that the variation in net multiplication of salmonellae in *Ity*^r and *Ity*^s macrophages is in fact a reflection of the extent of microbicidal activity. Furthermore, the recent reports by STACH et al. (1984) and CROCKER et al. (1984) directly demonstrated that the resident murine macrophage is the cell which expresses the *Bcg* and *Lsh* gene(s), respectively, as one would anticipate if *Bcg*, *Lsh*, and *Ity* are indeed the same gene.

2.3 *Lps*^d

Mice which are genotypically *Ity*^r can behave as if they are highly salmonella susceptible if they are also homozygous for the defective allele of the endotoxin response gene located on mouse chromosome 4. The C3H/HeJ mouse is such an endotoxin-hyporesponsive (SULTZER 1968), salmonella-susceptible animal (ROBSON and VAS 1972; VON JENEY et al. 1977; O'BRIEN et al. 1980c), and these two traits have been linked in a backcross study performed in our laboratory (O'BRIEN et al. 1980c). Thus, the *Lps*^d allele, or a gene closely linked to it, confers salmonella susceptibility on C3H/HeJ mice. Because virtually all C3H/HeJ cell types express LPS hyposensitivity (reviewed by VOGEL et al. 1981), the analysis of the mechanism of salmonella susceptibility in C3H/HeJ mice is complex. Nonetheless, we have demonstrated that C3H/HeJ mice die early (≤ 10 days) after parenteral infection and are unable to restrict the net multiplication of *S. typhimurium* in their spleens (O'BRIEN et al. 1980c). We have also obtained suggestive evidence that C3H/HeJ macrophages are not activated by the LPS on the surface of the bacterium and have postulated that such unstimulated phagocytes are unable adequately to restrict salmonella growth (O'BRIEN et al. 1982a). In another study (MACVITTIE et al. 1982) we observed that the influx of macrophages to the peritoneum of C3H/HeJ mice challenged intraperitoneally with *S. typhimurium* was only 1.7 times greater than basal level by 48–72 h of infection, whereas the macrophage influx for control mice was 3–4 times basal level. Taken together, these findings suggest that *S. typhimurium* fails adequately to "trigger" macrophages either to kill salmonellae or "call" other macrophages, either directly or through other cell types, into the site of infection. In vivo and in vitro studies are now in progress to assess the validity of these postulates.

2.4 C3HeB/FeJ

Mice of the C3HeB/FeJ strain are endotoxin responsive, i.e., homozygous for the normal *Lps*^n allele. However, EISENSTEIN et al. (1980, 1982) reported that these animals, like C3H/HeJ mice, are highly susceptible to *S. typhimurium* infection. These investigators suggested (EISENSTEIN et al. 1982) that the susceptibility of C3HeB/FeJ and C3H/HeJ mice to salmonellae might be due to the same gene and that such a gene must be distinct from the *Lps* locus. In a recent investigation (O'BRIEN and ROSENSTREICH 1983), we examined the relationship between the gene(s) which confers susceptibility to *S. typhimurium* on C3HeB/FeJ mice and other known *S. typhimurium* response genes. We tested F_1 hybrids derived from crosses of C3HeB/FeJ to mice which express other salmonella-susceptibility genes, and we found in all cases that gene complementation occurred. Consequently, we concluded that the susceptibility gene(s) of the C3HeB/FeJ mice is not Ity^s and is not the gene which confers salmonella sensitivity on C3H/HeJ mice (presumed to be Lps^d or an allele closely linked to it). Moreover, the C3HeB/FeJ susceptibility gene is also clearly different from *Ity*-2, a gene present in C57L mice that will be described below. Therefore, the susceptibility of C3HeB/FeJ mice to *Salmonella typhimurium* is regulated by a locus distinct from known salmonella response genes. The mechanism of C3H:HeB/FeJ salmonella susceptibility was not investigated, but we did observe that the animals died early in the course of salmonellosis.

3 Late-Stage Salmonella Response Genes

3.1 *xid*

CBA/N mice and F_1 male mice derived from CBA/N female parents carry an X-linked recessive allele designated *xid* for X-linked immunodeficiency. These animals express a variety of B-lymphocyte functional abnormalities that include poor antibody responses to some T-independent and T-dependent antigens, poor splenic B-cell proliferation responses to certain T-independent antigens, and low levels of serum IgM and IgG3. Most of the macrophage and T-cell functions of *xid* mice appear to be normal (SCHER 1981). The influence of *xid* expression on immunological responses of mice is reviewed in detail in the chapter in this volume by Wicker and Scher.

We evaluated the effect of the *xid*-conferred B-cell dysfunction on the response of mice to *S. typhimurium* (O'BRIEN et al. 1979b). The findings indicated that CBA/N mice, in contrast to immunologically normal, syngeneic CBA/Ca controls are *S. typhimurium* susceptible and that susceptibility is X-linked. Moreover, data from a backcross linkage analysis strongly suggested that the *S. typhimurium* susceptibility of CBA/N mice is in fact due to *xid* expression. In that same study, we noted that *xid* mice tend to die late in the course of murine typhoid.

The mechanism of *xid*-conferred susceptibility to murine typhoid was also evaluated in our laboratory (O'BRIEN et al. 1981). The four major observations made in that investigation are summarized below. First, salmonella-susceptible *xid* mice were phenotypically distinct from salmonella-susceptible Ity^s and Lps^d animals; *xid* mice were able to control the early net replication of *S. typhimurium* in their RES. Second, the delayed hypersensitivity responses of immune-defective F_1 male mice and their immunologically normal F_1 female littermates to salmonellae were similar. Third, the initiation of the IgG anti-salmonella antibody response in *xid* mice was delayed, and the magnitude of this humoral response was diminished following inoculation with killed *S. typhimurium* when compared with normal control. Fourth, immune-defective F_1 male mice could be protected from an otherwise lethal challenge with *S. typhimurium* when they were pretreated with salmonella-immune F_1 female serum. Further, the protective factor(s) in the F_1 female serum was removed by absorption with *S. typhimurium* and was contained in the gamma globulin portion of the serum. When viewed collectively, the data suggest that *xid* mice are susceptible to *S. typhimurium* infection because they fail to make an adequate protective antibody response that is apparently required for survival of infected mice late (> 10 days after parenteral inoculation) in the course of murine typhoid.

3.2 *Ity^r nu/nu* Mice

We examined the role of mature T cells in expression of early- and late-phase resistance of mice to *S. typhimurium* (O'BRIEN and METCALF 1982). For these experiments, we followed the net replication of *S. typhimurium* in the spleens and livers of nude *Ity^r* mice (CD-1 *nu/nu*) and their *nu/+* littermates with time. We found no significant difference in the growth patterns of salmonellae in the RES organs of T-cell-deficient and control animals for up to 13 days after intravenous challenge. Thereafter, net salmonella multiplication was greater in tissues of *nu/nu* than in the RES of *nu/+* mice. These findings clearly showed that mature T cells are not required for the containment of *S. typhimurium* growth during the early stages of murine typhoid, but that such cells are necessary for the ultimate survival of the salmonella-infected mice. The T cells may be required for protective anti-*Salmonella* antibody formation or for the development of acquired cellular immunity or for both.

3.3 C57L and DBA/2 Mice

Several groups of investigators have noted that mice of the C57L and DBA/2 strains are able to control the initial net growth of *S. typhimurium* in splenic and hepatic tissues, i.e., they are *Ity^r* animals, but they ultimately die when infected intraperitoneally (i.p.) with *S. typhimurium* (O'BRIEN et al. 1980b, 1982b; BENJAMIN and BRILES 1981; PLANT 1983). To assess the importance of route of challenge in the expression of this late-phase susceptibility to *S. typhimurium*, we performed a series of lethal dose studies with mice of the C57L/J, DBA/2J, C57BL/6 (*Ity^s*), and SWR/J (highly salmonella resistant, *Ity^r* animals) strains (O'BRIEN et al. 1984). We observed that the late-phase susceptibility of both the C57L and DBA/2 strains to murine typhoid was evident not only when mice were challenged i.p. but also subcutaneously, intravenously, and orally. We then analyzed the genetic basis for this susceptibility by tests of several F_1, F_2, and backcross hybrids derived from different matings between C57L/J, DBA/2, and SWR/J animals. Our results indicated that an autosomal recessive gene was primarily responsible for the deaths of salmonella-challenged C57L/J and DBA/2J mice, although additional minor genes appeared to influence the survival of these mice. We surmised that the major gene is distinct from all other previously defined *S. typhimurium* response genes. The chromosomal location of this late-acting gene has not as yet been determined; we found no linkage between expression of the susceptible phenotype and selected markers on chromosome 1, 2, 4, 5, or 7. How expression of the gene mediates salmonella susceptibility has not been determined. The finding that deaths of C57L/J and DBA/2J mice occur at a time in the course of the disease when salmonella-specific immune mechanisms (e.g., antibody formation) develop in mice that survive salmonellosis suggests that the defect conferred by expression of the late-phase gene may reflect a deficiency in acquired resistance to *S. typhimurium*. In support of such a contention, Plant has demonstrated that DBA/2 mice produce low levels of antibody to the *S. typhimurium* 0 antigen when compared with some other murine strains. Alternatively, the C57L and DBA/2 acquired immune deficiency may be one of cell-mediated immunity.

4 Conclusions

An inbred mouse will die following parenteral infection with $\geq$ ten bacteria of a highly mouse-virulent *S. typhimurium* strain if the animal is homozygous for any of the following alleles: Ity^s, Lps^d, C3HeB/FeJ gene(s), xid, Ity-2^s, or the C57L/DBA/2 gene(s). Mice that are homozygous for one or more of the first three alleles listed usually die within 10 days of challenge, whereas the latter three alleles confer late-stage susceptibility. For some of the alleles, the particular cell type responsible for the expression of salmonella sensitivity has been defined: the macrophage for Ity, the B cell for xid, and the T cell for nude mice. The precise mechanism by which salmonella susceptibility is conferred by the expression of any of the aforementioned alleles has not been established. Nonetheless, as more information accrues, we eventually hope to identify the products of at least some of these host genes that modulate resistance to murine typhoid.

Acknowledgments. This work was supported by NIH grant AI 17754-04 and USUHS protocol number R07313. I thank Mrs. Irmgard Dinger and Mrs. Sylvia Goldstein for secretarial assistance.

The opinions or assertions contained herein are the private views of the authors and should not be construed as official or as necessarily reflecting the views of the Uniformed Services University of the Health Sciences or Department of Defense. There is no objection to its presentation and/or publication.

References

Benjamin WH Jr, Briles DE (1981) Genetic studies with a mouse strain BRVR that is highly susceptible to both *Salmonella typhimurium* and *Listeria monocytogenes*. In: Annu Meet Am Soc Microbiol E51, p 63

Blumenstock E, Jann K (1981) Natural resistance of mice to *Salmonella typhimurium*; bactericidal activity and chemiluminescence response of murine peritoneal macrophages. J Gen Microbiol 125:173–183

Bradley DJ, Taylor BA, Blackwell J, Evans EP, Freeman J (1979) Regulation of *Leishmania* populations within the host, III. Mapping of the locus controlling susceptibility to visceral leishmaniasis in the mouse. Clin Exp Immunol 37:7–14

Brown IN, Glynn AA, Plant J (1982) Inbred mouse strain resistance to *Mycobacterium lepraemurium* follows the *Ity/Lsh* pattern. Immunology 47:149–156

Carter PB, Collins FM (1974a) Growth of typhoid and paratyphoid bacilli in intravenously infected mice. Infect Immun 10:816–822

Carter PB, Collins FM (1974b) The route of enteric infection in normal mice. J Exp Med 139:1189–1203

Crocker PR, Blackwell JM, Bradley DJ (1984) Expression of the natural resistance gene *Lsh* in resident liver macrophages. Infect Immun 43:1033–1040

Eisenstein TK, Deakins LW, Sultzer BM (1980) The C3HeB/FeJ mouse, a strain in the C3H lineage which separates salmonella susceptibility and immunizability from mitogenic responsiveness to lipopolysaccharide. In: Skamene E, Kongshaven PAL, Landy M (eds) Genetic control of natural resistance to infection and malignancy. Academic, New York, pp 115–120

Eisenstein TK, Deakins LW, Killar L, Saluk PH, Sultzer BM (1982) Dissociation of innate susceptibility to *Salmonella* infection and endotoxin responsiveness in C3HeB/FeJ mice and other strains in the C3H lineage. Infect Immun 36:696–703

Gerichter CB (1960) The dissemination of *Salmonella typhi*, *S. paratyphi* A, and *S. paratyphi* B through the organs of the white mouse by oral infection. J Hyg 58:307–319

Hohman AW (1978) Intestinal colonization and virulence of *Salmonella* in mice. Infect Immun 22:763–770

Hormaeche CE (1979a) Natural resistance to *Salmonella typhimurium* in different inbred mouse strains. Immunology 37:311–318

Hormaeche CE (1979b) Genetics of natural resistance to salmonellae in mice. Immunology 37:319–327

Hornick RB, Greisman SE, Woodward TE, Dupont HL, Dawkins AT, Snyder MJ (1970) Typhoid fever: pathogenesis and immunological control. N Engl J Med 282:686–691

Kessel RW, Monaco L, Marchisio MA (1963) The specificity of the cytotoxic action of silica – a study in vitro. Br J Exp Pathol 44:351–364

Lissner CR, Swanson RN, O'Brien AD (1983) Genetic control of the innate resistance of mice to *Salmonella typhimurium*: expression of the *Ity* gene in peritoneal and splenic macrophages in vitro. J Immunol 131:3006–3013

MacVittie TJ, O'Brien AD, Walker RI, Weinberg SR (1982) Inflammatory response of LPS-hyporesponsive and LPS-responsive mice to challenge with gram-negative bacteria, *Salmonella typhimurium*, and *Klebsiella pneumoniae*. In: Normann SJ, Sorkin E (eds) Macrophages and natural killer cells. Plenum, New York, pp 325–334

Maier T, Öels HC (1972) Role of the macrophage in natural resistance to salmonellosis in mice. Infect Immun 6:438–443

Nungester WJ, Jourdanais LF, Wolf AA (1936) The effect of mucin on infections by bacteria. J Infect Dis 59:11–21

O'Brien AD (1982) Innate resistance of mice to *Salmonella typhi* infection. Infect Immun 38:948–952

O'Brien AD, Metcalf ES (1982) Control of early susceptibility to *Salmonella typhimurium* growth in innately salmonella-resistant mice does not require functional T lymphocytes. J Immunol 129:1349–1351

O'Brien AD, Rosenstreich DL (1983) Genetic control of the susceptibility of C3HeB/FeJ mice to *Salmonella typhimurium* is regulated by a locus distinct from known salmonella response genes. J Immunol 131:2613–2615

O'Brien AD, Scher I, Formal SB (1979a) Effect of silica on the innate resistance of inbred mice to *Salmonella typhimurium* infection. Infect Immun 25:513–520

O'Brien AD, Scher I, Campbell GH, MacDermott RP, Formal SB (1979b) Susceptibility of CBA/N mice to infection with *Salmonella typhimurium*: influence of the X-linked gene controlling B lymphocyte function. J Immunol 123:720–724

O'Brien AD, Rosenstreich DL, Metcalf ES, Scher I (1980a) Differential sensitivity of inbred mice to *Salmonella typhimurium*: a model for genetic regulation of innate resistance to bacterial infection. In: Skamene E, Kongshaven PAL, Landy M (eds) Genetic control of natural resistance to infection and malignancy. Academic, New York, pp 101–104

O'Brien AD, Rosenstreich DL, Taylor BA (1980b) Control of natural resistance to *Salmonella typhimurium* and *Leishmania donovani* in mice by closely linked but distinct genetic loci. Nature 287:440–442

O'Brien AD, Rosenstreich DL, Scher I, Campbell GH, MacDermott RP, Formal SB (1980c) Genetic control of susceptibility to *Salmonella typhimurium* infection in mice: role of the *Lps* gene. J Immunol 124:20–24

O'Brien AD, Scher I, Metcalf ES (1981) Genetically-conferred defect in anti-*Salmonella* antibody formation renders CBA/N mice innately susceptible to *Salmonella typhimurium* infection. J Immunol 126:1368–1372

O'Brien AD, Metcalf ES, Rosenstreich DL (1982a) Defect in macrophage effector function confers *Salmonella typhimurium* susceptibility on C3H/HeJ mice. Cell Immunol 67:325–333

O'Brien AD, Rosenstreich DL, Scher I (1982b) Genetic control of murine resistance to *Salmonella typhimurium* infection. In: Friedman H, Klein TW, Szentivanyi A (eds) Immunomodulation by bacteria and their products. Plenum, New York, pp 37–48

O'Brien AD, Rosenstreich DL, Metcalf ES (1983) Host genes that influence the pathogenesis of murine typhoid. In: Keusch G, Wadstrom T (eds) Experimental bacterial and parasitic infections. Elsevier, New York, pp 31–38

O'Brien AD, Taylor BA, Rosenstreich DL (1984) Genetic control of natural resistance to *Salmonella typhimurium* in mice during the late phase of infection. J Immunol 133:3313–3318

Plant JE (1983) Relevance of the route of injection in the mouse model for *Salmonella typhimurium*. In: Keusch G, Wadstrom T (eds) Experimental bacterial and parasitic infections. Elsevier, New-York, pp 39–49

Plant J, Glynn AA (1976) Genetics of resistance to infection with *Salmonella typhimurium* in mice. J Infect Dis 133:72–78

Plant J, Glynn AA (1977) Mouse News Letter 57:38

Plant J, Glynn AA (1979) Locating *Salmonella* resistance gene on mouse chromosome 1. Clin Exp Immunol 37:1–6

Plant JE, Blackwell JM, O'Brien AD, Bradley DJ, Glynn AA (1982) Are *Lsh* and *Ity* at one locus on mouse chromosome 1? Nature 297:510–511

Potter M, O'Brien AD, Skamene E, Gros P, Forget A, Kongshaven PAL, Wax JS (1983) A BALB/c congenic strain of mice that carries a genetic locus (*Ity^r*) controlling resistance to intracellular parasites. Infect Immun 40:1234–1235

Powell CJ Jr, DeSett CR, Lowenthal JP, Berman S (1980) The effect of adding iron to mucin on the enhancement of virulence for mice of *Salmonella typhi* strain TY 2. J Biol Stand 8:79–85

Robson HG, Vas SI (1972) Resistance of inbred mice to *Salmonella typhimurium*. J Infect Dis 126:378–386

Rosenstreich DL, Weinblatt AC, O'Brien AD (1982) Genetic control of resistance to infection in mice. CRC Crit Rev Immunol 3:263–330

Scher I (1981) B-lymphocyte development and heterogeneity: analysis with the immune-defective CBA/N mouse strain. In: Gershwin E, Merchant B (eds) Immunologic defects in laboratory animals, vol 1. Plenum, New York, pp 163–190

Skamene EP, Gros A, Forget PAL, Kongshavn C, Charles St, Taylor BA (1982) Genetic regulation of resistance to intracellular pathogens. Nature 297:506–509

Spaun J (1964) Studies on the influence of the route of immunization in the active mouse protection test with intraperitoneal challenge for potency assay to typhoid vaccines. Bull WHO:31:793–798

Stach JL, Gros P, Forget A, Skamene E (1984) Phenotypic expression of genetically-controlled natural resistance to *Mycobacterium bovis* (BCG). J Immunol 132:888–892

Stuart AE (1970) Cellular immunity in bacterial infections. In: The reticuloendothelial system. Livingstone, Edinburgh, pp 122–162

Sultzer BM (1968) Genetic control of leukocyte responses to endotoxin. Nature 219:1253–1254

Swanson RN, O'Brien AD (1983) Genetic control of the resistance of mice to *Salmonella typhimurium*: *Ity* gene is expressed in vivo by 24 hours after infection. J Immunol 131:3014–3020

Taylor BA, O'Brien AD (1982) Position on mouse chromosome 1 of a gene that controls resistance to *Salmonella typhimurium*. Infect Immunol 36:1257–1260

Vogel SN, Weinblatt AC, Rosenstreich DL (1981) Inherent macrophage defects in mice. In: Gershwin ME, Merchant B (eds) Immunological defects in laboratory animals, vol 1. Plenum, New York, pp 327–356

Von Jeney N, Gunther E, Jann K (1977) Mitogenic stimulation of murine spleen cells: relation of susceptibility to *Salmonella* infection. Infect Immun 15:26–33

Genetic Control of Resistance to Mycobacterial Infection

E. Skamene

1 The Role of Genetic Factors in Mycobacterial Infections of Man

The idea of heredity in tuberculosis was widely accepted even prior to the discovery of the tubercle bacillus in 1882. There are several lines of evidence which strongly suggest that host genes controlling the innate resistance or susceptibility to this disease do exist in humans (SCHWEITZER 1961). The first line of evidence rests with the changing patterns of resistance in ethnic populations and other identified groups exposed to the mycobacteria over generations. It has been well established that initial exposure of populations to tuberculosis is associated with high mortality rates which decrease significantly with subsequent generations. On genetic principles, exposure over many generations should lead to death of susceptible individuals with survival of those who are genetically resistant. When tuberculosis was first introduced into the Qu'Appelle Valley

Division of Clinical Immunology and Allergy, Montreal General Hospital, 1650 Cedar Avenue, Montreal, Quebec, H3G 1A4, Canada

Current Topics in Microbiology and Immunology, Vol. 124
© Springer-Verlag Berlin·Heidelberg 1986

Indian Reservation in Saskatchewan the annual death rate from this disease reached 10% and over one-half of Indian families were eliminated in the first three generations. After 40 years and three generations, most susceptible individuals apparently died and the annual death rate had been reduced to 0.2% (MOTULSKY 1960). The presently high resistance of Western populations to tuberculosis is genetically conditioned through natural selection during long contact with the disease. The decline in tuberculosis mortality had begun before discovery of the tuberculosis organism and before medical measures were taken and probably is partially due to selective mortality of susceptible members of the population. These trends, although very suggestive of the importance of genetic determinants in tuberculosis, cannot by themselves be considered fullproof evidence since the important role of exposure has clearly been demonstrated in this disease.

Family studies, namely those concerning the comparisons of mono- and dizygotic twins as well as long-term epidemiologic studies of familial aggregation, provide a more convincing evidence of genetic determination of host susceptibility in tuberculosis. All twin studies clearly demonstrated increased concordance for the disease among monozygotic twins as compared with dizygotic twins (COMSTOCK 1978).

In view of the suggestive nature of these human data, conclusive animal experiments on genetic resistance to tuberculosis are of special interest.

2   Inherited Resistance to Tuberculosis and BCG Infection in Rabbits

By artificial selection several teams of investigators have succeeded in raising strains of rabbits which are either highly susceptible or resistant to tuberculosis. In 1932, LURIE started to inbreed rabbits for resistance and susceptibility to tuberculosis using the granulomatous response in the lungs developing after quantitative inhalation of virulent human-type tubercle bacilli (H37Rv) as a typing criterion. Brother-sister mating of resistant and susceptible rabbits led to the development of several inbred families which expressed different degrees of resistance to tuberculosis (LURIE et al. 1952a). The animals of resistant lines exhibited two key characteristics: they were more resistant to the establishment of infection (i.e., resistance to attack) and they had higher resistance to the progress of tuberculosis (i.e., resistance to continued multiplication of bacilli). The resistance to attack was attributed to the innately enhanced phagocytic activity of alveolar macrophages resulting in effective trapping of inhaled bacilli in the lung with subsequent rapid inactivation within the alveolar macrophages. Resistance to the progress of tuberculosis, characterized by increased longevity, after virulent infection, was associated with the greater restriction of bacillary accumulation presumably caused by a more rapid maturation of epitheloid cells in resistant rabbits (LURIE et al. 1952b). Histological studies revealed that after the inhalation of bacilli the resistant rabbits developed 20 times fewer number of primary tubercles than the susceptible rabbits. These tubercles were smaller and more interstitial in character and contained fewer bacilli than the primary lesions of susceptible rabbits. Although the invasion of the bloodstream by tubercle bacilli occurred in both the resistant and susceptible rabbits, the resistant ones inactivated a much larger proportion of bacilli that seeded the spleen, kidneys, and lymph nodes than did their susceptible counterparts (LURIE and DANNENBERG 1965). The resistance to progress of tuberculosis was attributed to the development of acquired immunity since it was associated with a more rapid development of tuberculin hypersensitivity in resistant rabbits. The response to intradermal BCG vaccination was also distinct in the susceptible and resistant rabbits (LURIE et al. 1952c). At 5 weeks after the inhalation of human-type tubercle bacilli, the number of primary tubercles in resistant rabbits was reduced 80% by BCG vaccination; the number in susceptible rabbits was reduced by 15%. Thus, resistant rabbits, which needed it the least, were protected the most and susceptible rabbits, which needed it the most, were protected the least. As it will be discussed later, this was a crucial observation which must be taken into account when trying to identify factors explaining the variability of effectiveness of the BCG vaccination against tuberculosis in humans.

Lurie's and Dannenberg's studies with genetic resistance of inbred rabbits to mycobacterial infection raised an interesting issue, namely that both the innate resistance and the acquired immunity are inherited and appear to be superior in the rabbits of resistant strains (LURIE et al. 1952a). The interpretation which was offered to explain these data was that the genetic system(s) governing these responses control an intrinsic function (i.e., increased microbicidal activity) of the macrophages of resistant animals. The development of superior acquired immunity in resistant strains was considered to occur as a secondary phenomenon, superimposed on, and determined by, native resistance (LURIE and DANNENBERG 1965). In other words, native resistance is merely a tendency for a rapid development of acquired resistance. Native susceptibility is essentially a tendency for a tardy or ineffective development of acquired resistance. The validity of such conclusion could, however, only be tested if the biochemical mechanism responsible for the innate resistance could be defined and shown to regulate the acquired immunity or if the linkage analysis could be performed in order to establish that the innate resistance and acquired immunity are under the same genetic control. In the absence of these lines of evidence an alternative hypothesis may be entertained, namely, that the gene(s) governing the superior innate resistance and acquired immunity became fixed in the resistant families by chance and those traits then coexist as unrelated phenomena. The attempt at Mendelian analysis of the trait of resistance to tuberculosis in these inbred rabbits resulted in the data suggestive of a multigenic control (LURIE et al. 1952b), making any further attempts on linkage studies between innate resistance and acquired immunity a difficult task. Lurie's and Dannenberg's experiments clearly established the importance of genetic factors in mycobacterial infections and have become the classics in the field. Their conclusions withstood the test of time and are now being confirmed, extended, and explained in the models of mycobacterial infection in inbred strains of mice.

3   Genetic Control of Tuberculosis and BCG Infection in the Mouse

3.1   Earlier Studies on Intraspecies Variation in Resistance/ Susceptibility to Intravenous Infection

Strain variation in resistance to mycobacterial infections among mice has been noted since the mid-1940s. Early reports by DONOVICK et al. (1949) showed that survival time of various mice strains after intravenous infection with *M. tuberculosis* varied considerably. The C57BL, DBA/1, and CFl strains were the most susceptible while the Swiss Albino, C3H, CFN, and Strong A (A/St) strains were the most resistant.

The unusual susceptibility of the C57BL/6 strain was also reported by GRAY (1960, 1961). Mice of this strain, when infected intravenously, could not restrict the proliferation of the bacilli in their reticuloendothelial organs and succumbed early in the course of infection. SEVER and YOUMANS (1957) observed a similar growth pattern of the avirulent *M. tuberculosis* H37Ra strain, the attenuated R1Rv strain, and the *M. bovis* (BCG) strain in the spleen, liver, and lungs of A/St and C57BL/6 mice. Mice of the C57BL/6 strain were, however, more susceptible to infection with the virulent *M. tuberculosis* H37Rv than the A/St and the CF1 strain, as measured by median survival time after intravenous inoculation (YOUMANS and YOUMANS 1972). Using *M. tuberculosis* (Vallée) as an infectious agent, LYNCH et al. (1965) performed a Mendelian segregation analysis of the trait of resistance to tuberculosis in (C57BL/6 × Swiss)F$_1$ hybrids and backcross populations. Continuous backcrossing showed that the trait of innate resistance was controlled by a single dominant gene which, with all probability, was identical to the chromosome 1 *Bcg* gene, which will be discussed later on.

KALEDIN et al. (1977) while studying the potentiation of antitumor responses by BCG also reported variation in the relative susceptibility of mouse strains to the infection with the BCG vaccine. They observed that, 4 weeks after the intravenous inoculation of 0.05 mg BCG, the number of viable bacilli recovered from the spleens of C57BL/6 animals exceeded 20 times that recovered from C3Hf mice. Employing a lower infective dose of BCG (0.0025 mg), these authors observed an even higher difference (100-fold) between the numbers of viable BCG bacilli recovered from the spleens of C57BL/6 mice when compared with C57BR mice at 14 days after the infection. The authors suggested that C57BL/6 mice were highly susceptible while C3Hf and C57BR mice were resistant to BCG infection. In their studies on genetic restriction of BCG-induced protection against murine leprosy, LAGRANGE and HURTREL (1979) showed that the BCG bacilli, when injected intravenously, multiplied markedly in the spleens of C57BL/6 mice. On the other hand, no multiplication occurred in C3H mice and the bacilli were steadily eliminated from the spleen during the 5-week period.

An interesting observation, commonly noted in all these earlier studies dealing with the relative susceptibility of different mouse strains to intravenous infection with *M. tuberculosis* and *M. bovis* (BCG), was the marked susceptibility of C57BL/6 mice. The innate susceptibility of this strain to mycobacterial infection seemed to be expressed very early in the course of host-parasite interaction.

3.2 The *Bcg* Gene: Genetics and Phenotypic Expression

3.2.1 Inbred Strain Survey

Having noted the earlier studies suggesting that the susceptibility to BCG infection is under genetic control, we have embarked on a formal genetic analysis of this trait. The experimental conditions chosen in our experiments were such that an optimal BCG multiplication could occur in the mice of susceptible strains after intravenous infection. Great care was taken to use only a clump-free, dispersed, monocellular suspension of bacilli for injection (FORGET et al. 1981). Furthermore, considering the fact that BCG multiplication in vivo is inversely related to the dose of injected inoculum (BENOIT and PANISSET 1963; LEFFORD 1970), a small dose ($\simeq 10^4$ viable BCG units) was used to infect mice for the genetic studies. Under these conditions, the intravenously injected BCG multiplies rapidly in the spleens and livers of susceptible mice in the first 3–4 weeks of infection. No multiplication occurs in that time period in the reticuloendothelial organs of resistant mice. Three weeks after the infection, the mycobacterial burden detectable in the spleens of susceptible animals is about 100–1000 times higher than that present in the spleens of resistant animals. This time point in the kinetics of host-parasite interaction was chosen as a typing criterion for resistance/susceptibility. Mice of 25 commonly used inbred strains, when screened for this trait, segregated in the two sharply distinct categories. About one-half of all the strains typed as resistant and these included A/J, AKR/J, C57Br/J, C3H/HeCr, C3H/HeJ, C3HeB/FeJ, CBA/J, CBA/N, DBA/2J, C57BL/6J, and 129/J. Among the strains classified as susceptible to BCG infection were, as expected, the C57BL/6-derived mice (C57BL/6J, C57BL/10J, B10.A, and B10.D2) as well as DBA/1J, BALB/cJ, BALB/cAnNCr, CE/J, SWV, BSVR, and BSVS. An important observation in these studies was the great homogeneity of typing results obtained in animals of the same strain (either susceptible or resistant) and the great homogeneity observed between strains of the same phenotype. The clear segregation into two distinct groups, without the presence of any "overlapping" strain (of intermediate resistance) suggested that the differences in response to BCG infection among inbred mice may be controlled by a simple genetic system.

3.2.2 Segregation Analysis of the Mode of Inheritance of the Trait of BCG Resistance/Susceptibility

The F_1 hybrids derived from the cross of resistant and susceptible mice were uniformly resistant, implying that the resistance was dominant over susceptibility. F_1 hybrids derived from the cross of susceptible progenitors were uniformly susceptible. Since no gene complementation was observed in these studies, it could be concluded that the genetic control of susceptibility in all the different BCG-susceptible strains was exerted by the same gene(s). Among the progeny of F_1 hybrids backcrossed to the susceptible progenitor, approximately 50% of animals were found to be resistant while the resistant F_1 hybrids backcrossed to the resistant progenitors were all resistant. Approximately 75% animals of the F_2 generation were resistant. These ratios of resistant to susceptible individuals of the hybrid and backcross generations are compatible with the hypothesis that the trait of resistance to BCG infection is under the control of a single, dominant, autosomal gene (no influence of the sex on the level of resistance was observed in any groups). Similar conclusions were derived from backcross analysis performed in different pairs of resistant/susceptible strains. The gene which was identified as the regulator of this early phase of resistance to BCG was given the provisional designation *Bcg* (GROS et al. 1981).

It exists in one of the two allelic forms, Bcg^r (resistant) or Bcg^s (susceptible). A successive backcrossing of the trait of BCG resistance onto a background of a susceptbile progenitor led to a continuing segregation of the progeny in a 50:50 (resistant:susceptible) ratio, thus confirming the monogenic control. In fact, this procedure, performed now for over 20 generations, led to the construction of a *Bcg* congenic line in which a Bcg^r allele donated by the A/J strain was transferred onto an innately susceptible C57BL/10 background (FORGET et al. 1986, unpublished observation). A success in the establishment of a congenic lines represents the formal proof of the hypothesis that the innate resistance to BCG infection is controlled by a single gene.

3.2.3   Mapping of the Bcg Gene to Chromosome 1

It was considered that the mapping of the locus carrying the *Bcg* gene would be informative in two ways: its map location might correspond to the position of an already mapped locus whose phenotypic expression is well characterized (ex: enzyme). This finding might give information on the product of the *Bcg* gene. Another expected outcome of the mapping studies might be the identification of another gene which has already been mapped, in the vicinity of the *Bcg* locus. The product of such a known gene could be used as a marker in the typing of BCG resistance/susceptibility in segregating population and it can also be of help in the construction of the *Bcg*-congenic lines.

The mapping of the *Bcg* gene was achieved by typing the recombinant inbred (RI) strains for the trait of resistance/susceptibility to BCG. The BXH and BXD RI strains used for this study were derived from the pairs of progenitors expressing the allelic forms of *Bcg* gene, namely the C57BL/6J (B, Bcg^s) and C3H/HeJ (H, Bcg^r or DBA/2J (D, Bcg^r). The strain distribution pattern of Bcg^s and Bcg^r alleles in 36 RI strains was found to be very similar to the strain distribution pattern of two isoenzyme loci, namely *Idh-1* (isocitrate dehydrogenase) and *Pep-3* (peptidase 3), located on the proximal (centromeric) part of chromosome 1 (SKAMENE 1983). The recombination frequency between the *Bcg*, *Idh-1*, and *Pep-3* markers established the gene order in that region of first chromosome as *Idh-1 – Bcg – Pep-3*.

An even more interesting outcome of the mapping studies was the finding of identity of the strain distribution pattern of Bcg^r and Bcg^s alleles with that of the Ity^r and Ity^s alleles controlling the resistance/susceptibility of mice to *Salmonella typhimurium* (O'Brien, this volume) and of Lsh^r/Lsh^s alleles regulating the innate resistance/susceptibility to *Leishmania donovani* (Bradley, this volume). An identity of the *Bcg*, *Ity*, and *Lsh* genes, predicted on the basis of RI strain mapping, was confirmed by a backcross linkage analysis (SKAMENE et al. 1982). It should be pointed out that the conclusion about identity of the *Bcg*, *Ity*, and *Lsh* genes is only a tentative one at the moment, awaiting confirmation by the molecular genetics approach. However, three factors are strongly in favor of gene identity (as opposed to a close *Bcg-Ity-Lsh* linkage in a chromosomal complex): firstly, no proven recombination between *Bcg*, *Ity*, and *Lsh* genes was found after several years of backcross testing. Secondly, several congenic lines have now been made independently in three laboratories using the *Lsh*, *Ity*, or *Bcg* typing as a selective pressure; in all cases tested so far the allelic distribution of all three phenotypes remained identical (POTTER et al. 1983). Finally, studies on the phenotypic expression of *Ity*, *Lsh*, and *Bcg* genes all led to the same conclusion regarding the mechanisms regulated by this genetic control.

3.2.4 Phenotypic Expression of the Bcg Gene

When considering the host defense mechanism underlying the *Bcg* gene-controlled innate resistance to infection with *M. bovis* BCG, it is necessary to remain within the framework of the experimental model that defines the phenotype: the *Bcg* gene controls the proliferation of the mycobacteria in vivo after intravenous infection with a relatively low dose of BCG. Even before embarking on the series of experiments designed to evaluate various parameters of host resistance to mycobacterial infection, it was obvious that the *Bcg* is unlikely to control a specific, acquired immunological response to the BCG bacilli. *Leishmania donovani* and *Salmonella typhimurium* are antigenically distinct from each other and from mycobacteria; yet the *Bcg* (*Lsh*, *Ity*) gene regulates the host response to all the three pathogens. The phenotype of resistance to BCG can be demonstrated early on in the course of infection. The clearance of intravenously injected BCG bacilli from the bloodstream and their uptake in the reticuloendothelial tissues (mainly liver and spleen) is similar in both the Bcg^r and Bcg^s host. However, within 24–48 h the genetic advantage of Bcg^r mice becomes obvious in that the numbers of viable BCG bacilli recovered from their spleens are significantly lower than those which are found in the Bcg^s spleens. This difference becomes more significant as the infection progresses up to its peak 2–3 weeks later. A similar pattern of the kinetics of bacillary burden in the Bcg^s and Bcg^r strains can be demonstrated with a variety of other BCG substrains (DENIS et al. 1986) as well as with other mycobacteria such as *M. lepraemurium* (BROWN et al. 1982; SKAMENE et al. 1984), *M. intracellulare* (GOTO et al. 1984), and several other atypical mycobacteria. The differences between Bcg^s and Bcg^r mice in the number of viable mycobacterial bacilli which can be found after i.v. infection are best demonstrable in the spleen but are seen as well in other organs which are rich in the reticuloendothelial tissues such as liver and lungs. Given the fact that all the pathogens which are restricted in their in vivo growth by the *Bcg* gene are intracellular parasites of the macrophage, one could intuitively consider the macrophage to be the best candidate for the cell expressing the phenotype of genetic resistance in these models. The inflammatory responses to the seeding of bacilli in the tissues early on in the course of infection seem to be identical in the spleens and livers of Bcg^r and Bcg^s animals. Similarly, when bacilli are introduced intraperitoneally (a route which also leads to discrimination, albeit less perfect, between the Bcg^r and Bcg^s phenotype) the quantity of inflammatory macrophages and other cells of the peritoneal exudates is identical in both types of animals. The Bcg^r animals remain resistant to BCG infection even after a 900-R dose of radiation, suggesting that the cell responsible for genetic resistance is mature and does not require division to express the resistant phenotype. This cell population originates in the bone marrow as demonstrated in the chimera experiments: The Bcg^s animals were lethally irradiated and subsequently rescued by the transplantation of bone marrow derived from their Bcg^r-congenic counterparts. After the establishment of chimerism (10 weeks after transplantation), they were infected with BCG and found to be as resistant as the donors of the bone marrow. Similarly, reciprocal $Bcg^s \rightarrow Bcg^r$ chimeras became phenotypically susceptible to BCG.

The use of mutants affecting the T-cell function, B-cell function, and NK cell function revealed that these cell compartments are not involved in the genetic resistance to BCG infection. This series of experiments thus strengthened the initial impression that the mature macrophage of the spleen, liver, and lungs is responsible for the ability of the Bcg^r host to restrict the initial bacillary proliferation in vivo. This impression was further enhanced when it was demonstrated that the only maneuver which could transform the Bcg^r animals into phenotypically BCG-susceptible was chronic treatment with silica (GROS et al. 1983).

3.2.5 Resident Macrophage Expresses the Bcg Gene

All the in vivo evidence implicating the macrophage as the cell expressing the phenotype of genetic resistance to BCG has been circumstantial and a direct in vitro proof was needed to confirm that conclusion. The in vitro assays of the interaction between macrophages and mycobacteria are traditionally difficult to perform and evaluate because of the relatively slow growth rate of the bacilli within explanted macrophages and of the clumping of bacilli, leading to erroneous counts of viable bacilli under those conditions. In order to circumvent these problems, a radiometric method based on the incorporation of ^{3}H-uracil into proliferating bacilli was recently developed by ROOK and RAINBOW (1981). It was subsequently demonstrated using this method that both the Bcg^r and Bcg^s resident peritoneal macrophages inhibited the rate of ^{3}H-uracil incorporation into the BCG bacilli but that the inhibition by Bcg^r cells was significantly higher than that observed with Bcg^s macrophages and that such inhibition lasted even after the BCG bacilli were released from the intracellular milieu of the macrophage (STACH et al. 1984). Using different methodologic approaches it was concluded that the resident macrophage (explanted in vitro) also expresses the Ity^r phenotype (LISSNER et al. 1983) and the Lsh^r phenotype (CROCKER et al. 1984). There is no doubt, therefore, that the resident macrophage alone, without the need for any interaction with an other cellular compartment, is endowed with an intrinsic, genetically controlled, ability to limit intracellular proliferation of certain pathogens including the BCG.

3.2.6 Concept of Two Phases of Host Response to BCG Infection: Innate Resistance and Acquired Immunity

A bactericidal (or bacteriostatic) mechanism of the resident macrophages, controlled by the *Bcg* gene, constitutes a purely native mechanism of resistance against the infection with BCG and other mycobacterial pathogens. This conclusion was further enhanced by the observation that the parameters of acquired immune response to BCG were low or undetectable in some Bcg^r strains (such as A/J) in the early phase of host response to intravenous BCG infections, at the time that the bacillary growth in the spleens and livers was prevented. These animals thus resisted infection in the absence of granuloma formation or delayed-type hypersensitivity (DTH, footpad swelling) to BCG antigen. Other parameters of acquired immunity such as resistance to heterologous infection (*Listeria monocytogenes*) or to homologous challenge with higher doses of BCG were also not demonstrable. The conclusion derived from these studies was that the antigenic load of BCG present in these animals (less than 10^4 per spleen) was too low to stimulate any detectable T-cell immunity. The susceptible Bcg^s animals, on the other hand, when infected with low doses of BCG (10^4 live bacilli) allowed the inoculum to multiply in vivo 100–1000 times, up to the level of 10^6–10^7 bacilli per spleen and liver by the 3rd week of infection. From that point on, further BCG multiplication ceased and the animals started to clear the infection progressively. Associated with the elimination of BCG bacilli, the development of several parameters of acquired immunity (granuloma formation, DTH, resistance to heterologous and homologous challenge) was observed. The temporary association between the development of anti-BCG immunological responsiveness and the resistance against further growth of BCG in vivo which was demonstrated in these innately susceptible hosts suggested that another mechanism, namely acquired immunity, was responsible for the downturn in the bacillary load later on in the course of infection. The concept of two phases of responses to mycobacterial infection (innate resistance and acquired immunity), mechanistically distinct from each other but functionally interrelated in the context of host response, has thus been confirmed in the mouse model (PELLETIER et al. 1982). Further investigations of this phenomenon, performed in the congenic pairs of Bcg^r and Bcg^s mice, allowed us to conclude that the kinetics and the quality of acquired responses to BCG are directly related to the level of BCG load in the tissues, which, in turn, is the result of the infecting dose and the *Bcg* type of the host (BOURASSA et al. 1985). Bcg^r mice, infected with the low BCG dose, restrict the bacillary proliferation in the absence of detectable immunity. An equivalent dose of live BCG given to Bcg^s mice results in the accumulation of the load of bacilli which is high enough to stimulate the acquisition of protective immunity. A higher dose of

BCG (10^6 live bacilli) given to Bcg^r hosts leads to the development of protective immunity while the same dose, when used to infect the Bcg^s mice, results in the accumulation of such an antigenic overload that the phenomenon of suppression (demonstrable both in vivo and in vitro) appears. Under those conditions (infection with relatively high BCG numbers), the genetically resistant mice developed a superior immunity while the genetically susceptible mice became immunosuppressed.

3.3 Macrophage Mediated Genetic Resistance to BCG in Biozzi Mice

By selective breeding of outbred Swiss mice for quantitative antibody response to a complex immunogen (sheep red blood cells), BIOZZI et al. (1984) obtained two lines of mice which were high (HL) or low (LL) antibody producers to many other unrelated antigens including microorganisms, BCG being among them. These differences in antibody production were shown to be related to the segregation of polygenic characters determined by a group of about 10–12 independent loci. Several hypotheses have been offered to explain the mechanism of the HL and LL phenotypes. An attractive one is based on the observation that the breeding selected for macrophage function: macrophages from LL mice were shown to be endowed with a higher metabolic capacity to degrade antigens than those from HL mice and, as a result, presented the antigens poorly (WIENER and DANDIERI 1974). When injected with the BCG intravenously, mice of the two lines differed significantly in their ability to restrict bacillary multiplication. Although no differences were noted in the uptake of BCG by the spleens and liver (by 6 h of injection), the LL mice were at least ten times more resistant by the 3rd week of infection than the HL mice. Genetic resistance of the LL mice was not associated with the superior development of the cellular immunity to BCG. On the contrary, the genetically susceptible HL mice exhibited stronger DTH to tuberculin and much higher T-cell mitogenic response to BCG in vitro (LAGRANGE et al. 1979). Here again, the essential role of the macrophage in the genetic resistance to BCG infection was suggested (although not formally proved). Similarly to the model of Bcg gene-mediated genetic resistance, the acquired immunity was considered to be a secondary phenomenon developing as a response to higher multiplication of BCG which was allowed to occur in the HL line as a result of the lower intrinsic macrophage antimycobacterial activity. The interesting question which has not yet been answered experimentally is whether the Bcg gene might not be one of the genes fixed in the Biozzi mice during the selective breeding.

3.4 Genetic Control of Resistance to Subcutaneous Infection with BCG

In contrast to the striking interstrain differences in the control of mycobacterial multiplication in the reticuloendothelial organs after intravenous infection with BCG, it appears that the extent of local proliferation after subcutaneous BCG infection is rather similar in different inbred strains. However, genetically determined variation in the control of the BCG load which accumulates in the spleens after subcutaneous infection has been reported (NAKAMURA and TOKUNAGA 1978). Mice of SWM strains showed progressive decrease in the number of viable BCG recovered from the spleens between the 2nd and 3rd week after subcutaneous infection with 2×10^7 BCG. On the other hand, no BCG elimination was noted in the spleens of C3H mice. Since the genetic difference between C3H and SWM mice appeared rather late in the course of infection, one could assume that the mechanisms responsible for this effect operated at the level of acquired immunity. In support of this conclusion, it was found that the onset of resistance to BCG in SWM mice was characterized by splenomegaly and by the development of strong DTH to PPD. Susceptible C3H mice showed no increase in the spleen size and exhibited a cutaneous anergy. Segregation analysis suggested that the DTH response to BCG infection is controlled by a single gene, not linked to *H-2* or to *Bcg* loci. Studies on the phenotypic expression of this gene indicated that the high- and low-responder mice differed in the antigen-presenting ability of the splenic adherent cells characterized by a low antigen-specific blastogenesis and a low lymphokine production in vitro (NAKAMURA et al. 1982a). More specifically, I-J-positive macrophages of the low-responder (susceptible) C3H/He strain induced suppressor T cells when cultured with (C3H/He × SWM/Ms)F$_1$ T cells in vitro (NAKAMURA et al. 1982b). This line of experiments thus points out once again the importance of genetically-controlled macrophage function on the development of specific immune response and of the acquired immunity to BCG infection. It is of interest to note that some patients with recently diagnosed tuberculosis could be characterized as low responders to tuberculin. Low responsiveness was correlated with the presence of circulating adherent suppressor cells, probably monocytes. These findings seemed to have a genetic basis although the genotypes of these patients were not studied (ELLNER 1978).

3.5 Genetic Control of Immunological Responses to BCG

Although the role of specific immunological responsiveness in the models of genetic resistance or susceptibility to BCG infection is not clear, there is no question that various parameters of immune response to BCG (whether related to antimycobacterial resistance or not) *are* under genetic control. It is useful to keep these models in mind when considering the various levels of possible genetic regulation of host responses to BCG.

3.5.1 Genetic Control of Granuloma Response to Oil-Associated BCG Cell Wall Vaccine

Cell walls of BCG (BCG-CW) associated with minute oil droplets and dispersed in Tween-saline cause a remarkable increase in lung weight when injected intravenously into mice. Histological examinations revealed that the increase in lung weight resulted from granuloma formation. Inbred mice can be classified into three categories according to lung weights 1 month after BCG-CW vaccination: C57BL/6, SJL/J, and AKR/He are high responders whereas C3H/He, C3HeB/ FeJ, DBA/1J, and P/J are low responders. The responses of C57BL/10SnJ, DBA/2, BALB/c, and C3H/HeJ mice are intermediate. The segregation analysis revealed that the trait of lung granuloma formation with BCG-CW vaccine was under polygenic control, with a minor involvement of *H-2* locus (YAMAMOTO and KAKINUMA 1978). An interesting observation was the dissociation of BCG-CW-induced lung granuloma formation and footpad DTH to purified protein derivative (PPD) in a number of inbred strains tested.

Two strains were used as representative of high (C57BL/6) and low (C3H/ He) responders in studies dealing with the mechanisms responsible for these genetic differences (KAKINUMA et al. 1981; KATO and YAMAMOTO 1982; KAKI-MUMA et al. 1983). Bone marrow chimera studies revealed that the cell responsible for the high response originated in the bone marrow and cell transfer experiments demonstrated clearly that the macrophages were the cellular compartment expressing the low response of C3H/He mice. The mechanism proposed to explain the reduced granuloma formation in the low responder strain was the failure of effective antigen presentation by macrophages. Impaired antigen presentation resulted in the induction of suppressor T cells which were ultimately responsible for reduced granuloma formation. Interestingly, the immunological status of low-responder C3H/He mice could be manipulated by selection of route of immunization: BCG-CW given subcutaneously led to appropriate sensitization and to normal lung granuloma formation. Finally, evidence for a direct correlation between lung granuloma formation and antituberculous protection (resistance to live virulent challenge by aerosol) was provided: low-responder C3H mice were not protected while the high-responder C57BL/6 mice exhibited a high degree of protection.

3.5.2 Immunogenetics of BCG-Induced Anergy

An intravenous injection of dead BCG in an oil-in-saline emulsion produces diffuse interstitial and alveolar granulomatous lung disease as well as splenomegaly in certain inbred strains of mice (ALLEN et al. 1977; MOORE et al. 1981) that is under multigenic control (SCHRIER et al. 1982). The strain distribution pattern of high and low responders is similar to that reported for the trait of lung granuloma induced by the BCG-CW vaccine and it is possible that the genetic control of both traits is identical. Many of the inbred murine strains that develop the diffuse interstititial lung disease and splenomegaly are anergic with respect to responsiveness to mitogens in vitro as well as to unrelated and specific antigens in vivo (ALLEN et al. 1977; ALLEN and MOORE 1979; MOORE et al. 1981; SCHRIER et al. 1980). BCG-induced cutaneous anergy as evaluated by suppression of DTH to sheep erythrocytes is mediated by adherent, Thy-1$^-$, Ig$^-$ spleen cells. BCG-induced anergy (DTH to SRBC) is recessive, autosomal, unigenic, and controlled by a gene linked to the Igh-1 allotype on chromosome 12. This trait is, furthermore, influenced by genes linked to the H-2 complex (CALLIS et al. 1983).

3.5.3 BCG-Induced In Vivo Release of Lymphokines in Different Strains of Mice

Mice which are sensitized intravenously with either dead BCG in an oil-in-Tween emulsion or BCG-CW vaccine in an oil-in-Tween emulsion release lymphokines in vivo when challenged intravenously with an old tuberculin 3 weeks later. When the levels of migration inhibitory factor (MIF) or interferon (IF) are measured under such experimental conditions, the mouse strains segregate into high and low responders (NETA and SALVIN 1980). Again, the strain distribution pattern of high (C57BL/6, DBA/2, BALB/c) and low (C3H/He, AKR, CBA/Ca, NZB/W, DBA/1) response resembles that of lung granuloma formation discussed in the previous section. Also, similarly to the model of BCG-induced lung granuloma formation, the low-responder strain can be induced to produce lymphokines in vivo by a subcutaneous BCG-in-oil challenge.

4 Strategies of the Genetic Analysis of Host Response to BCG

Considering the complicated interaction between BCG and the host, it is not surprising to see that the host response to this infectious agent is regulated by multiple genes. However, by selecting a well-defined aspect of the host response for genetic studies, it is at times possible to dissect the action of individual genes and to attempt to piece together the multigenic control as a succession of events which are under the control of single genes. Resistance to intravenous infection with a low BCG dose may be used as an example of such an approach. If we measure the degree of BCG burden in the spleens and livers of mice infected 6 weeks previously with 10^4 live BCG bacilli, we see no difference in the numbers of BCG recovered from the organs of various inbred mouse strains and there is a temptation to conclude that there is really is no genetic control of the bacillary multiplication by the host. Furthermore, when measuring the different parameters of BCG-induced immunity such as granuloma formation, DTH, PPD-stimulated blast transformation, challenge to reinfection, and passive transfer of resistance, we can see striking differences between mice of various strains can be seen, leading to an (erroneous) conclusion that there is no relationship between immunity and resistance. However, when a careful kinetic study of the course of infection is performed it is clearly seen that there are different mechanisms responsible for the fact that various mouse strains all eventually reach the same level of BCG burden in their tissues (Fig. 1).

Mice of the resistant strain have low numbers of BCG bacilli in their organs throughout the course of infection. The mechanisms (macrophage mediated) of innate resistance never allowed the *establishment of infection* in these hosts. Significant bacillary multiplication never occurred and no additional mechanism

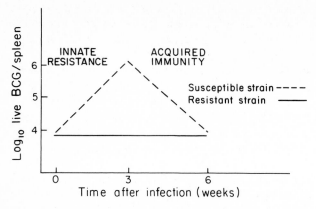

Fig. 1. Kinetics of splenic bacillary load after intravenous infection with 10^4 BCG

was required to keep the BCG numbers at the low levels. Mice of the susceptible strain also have a low number of BCG bacilli in their spleens by the 6th week of infection. However, one can easily appreciate that the mechanism of successful host defense is quite different from that operating in the resistant hosts. The mechanism of innate resistance, deficient in the susceptible hosts, allowed the establishment of infection by the 2nd–3rd week. Increasing antigenic load results in the development of acquired immunity, macrophage activation, and restriction of bacterial growth. The phase of innate resistance is under the control of the *Bcg* gene; one would intuitively suspect that the acquired immunity is under the control of immune response genes. Yet, the clear association between the H-2 (or HLA) complex and the resistance/susceptibility to BCG infection has not been established. This is mostly likely caused by the fact that the role of innate factors in the expression of resistance to BCG has not been appreciated. It may be predicted that the control of the acquired phase of response to BCG by the *Ir* genes will be demonstrated when congenic resistant mice of innately susceptible Bcg^s background (such as C57BL-derived, or BALB-derived) would be used for analysis. In fact, it has recently been shown that the different H-2 congenic strains constructed on a C57BL/10 background show genetic restriction of their DTH to PPD after BCG immunization (LAGRANGE et al. 1983).

The mechanisms of innate resistance to BCG may involve the regulation of antigen presentation by the macrophage and it is possible that the quality of immune response (i.e., protective versus suppressive) will be dictated by the *Bcg* genetic background and the dose of BCG used for infection. Experiments with different *Bcg* congenic lines which are presently available in several laboratories may well show that the often observed suppression of cellular immune responsiveness in mycobacterial infections occurs as a consequence of innate (i.e., macrophage-mediated) susceptibility. Similarly, the influence of genetic resistance/susceptibility on the success of BCG vaccination against infectious agents will need to be considered, both in animal models (CIVIL and MAHMOUD

1978) and in human trials (LAGRANGE et al. 1983). Similarly, the often-erratic response of mice to BCG-induced tumor immunoprophylaxis could probably be accounted for, to a certain degree, by the strain variation in the response to BCG itself (TOKUNAGA et al. 1978; KALEDIN et al. 1978).

The use of genetic analysis may help us, in the future, to answer some questions as to the mechanisms of protective acquired immunity to BCG infections. The body of data about the genetic control of various parameters of anti-BCG immunity is accumulating but, so far, most of the available evidence suggests a dissociation between phenomena such as DTH, granuloma formation, lymphokine production, etc., and the degree of protection against infection. The inflammatory reactions (DTH and granuloma formation) although developing in response to the established infection with the same kinetic fashion as the acquired resistance to infection may well be mediated by different cellular subpopulations and directed against different components of the mycobacteria. The identification of mouse strains which show genetic variation in one particular parameter of BCG-induced immune response should allow us to search for a functional association with the trait of acquired BCG resistance.

5 Conclusion

Experimental models discussed in this review clearly establish several levels of genetic control of host response to BCG infection. None of these models confirm identically to the course of human infection but they can all be used to explore some particular aspects of the protective mechanisms. In addition to these theoretical considerations about the usefulness of genetic models as a tool to study mechanisms of resistance, there is one major practical aspect of this approach. Mycobacterial infections (leprosy and tuberculosis) still represent a major health problem and the role of BCG vaccination continues to be hotly debated (BAILEY 1980). It has often been observed that, after vaccination with BCG, a certain percentage of vaccinees are insufficiently protected (MRC 1963). From the preventive point of view, this small proportion of the population represents the high-risk group. These people are probably the equivalent of Bcg^s mice in that their macrophages are defective in handling the bacilli. It is essential to be able to identify the individuals of this high-risk group, perhaps by the in vitro macrophage typing as developed in the mouse model, with the view of planning the appropriate strategies of their protection.

References

Allen EM, Moore VL (1979) Suppression of phytohemagglutinin and lipopolysaccharide responses in mouse spleen cells by Bacillus Calmette-Guérin. J Reticul Soc 26:349–356
Allen EM, Moore VL, Stevens JO (1977) Strain variation in BCG-induced chronic pulmonary inflammation in mice. I. Basic model and possible genetic control by non-H-2 genes. J Immunol 119:343–347

Bailey GVJ (1980) Trail of BCG vaccines in South India for tuberculosis prevention. Indian J Med Res 72[Suppl]:1–74

Benoit JC, Panisset M (1963) Survie et multiplication du BCG et du bacille tuberculeux chez la souris. IV. Survie et multiplication des souches filles de BCG en fonction du temps et de la dose. Acta Tubercol Pneumol Scand 43:125–136

Biozzi G, Mouton D, Stiffel C, Bouthillier Y (1984) Major role of macrophage in quantitative genetic regulation of immune responsiveness and anti-infection immunity. Adv Immunol 36:189–234

Bowrassa D, Forget A, Pelletier M, Skamene E, Turcotte R (1985) Cellular immune responses to *Mycobacterium bovis* (BCG) in genetically susceptible and resistant congenic mouse strains. Clin Exper Immunol 62:31–38

Brown IN, Glynn AA, Plant J (1982) Inbred mouse strain resistance to *Mycobacterium lepraemurium* follows the *Ity/Lsh* pattern. Immunology 47:149–156

Callis AH, Schrier DJ, David C, Moore VL (1983) Immunogenetics of BCG-induced anergy in mice. Control by Igh- and H-2-linked genes. Immunology 49:609–616

Civil RH, Mahmoud AAF (1978) Genetic differences in BCG-induced resistance to *Schistosoma mansoni* are not controlled by genes within the major histocompatibility complex of the mouse. J Immunol 120:1070–1072

Comstock GW (1978) Tuberculosis in twins: a re-analysis of the Prophit study. Am Rev Respir Dis 117:621–624

Crocker PR, Blackwell JM, Bradley DJ (1984) Expression of the natural resistance gene *Lsh* in resident liver macrophages. Infect Immun 43:1033–1040

Denis M, Forget A, Pelletier M, Turcotte R, Skamene E (1986) Control by the *Bcg* gene of early resistance in mice to infections with BCG substrains and atypical mycobacteria. Clin Exper Immunol (in press)

Donovick R, McKee CM, Jambor WP, Rake G (1949) Use of the mouse in standardized test for anti-tuberculous activity of compounds of natural or synthetic origin: choice of mouse strain. Am Rev Tubercol 60:109–120

Ellner JJ (1978) Suppressor adherent cells in human tuberculosis. J Immunol 121:2573–2579

Forget A, Skamene E, Gros P, Miailhe A-C, Turcotte R (1981) Differences in response among inbred mouse strains to infection with small doses of *Mycobacterium bovis* BCG. Infect Immun 32:42–47

Goto Y, Nakamura RM, Takahashi H, Tokunaga T (1984) Genetic control of resistance to *Mycobacterium intracellulare* infection in mice. Infect Immun 46:135–140, 1984

Gray DF (1960) Variations in natural resistance to tuberculosis. J Hyg 58:215–227

Gray DF (1961) The relative natural resistance of rats and mice to experimental pulmonary tuberculosis. J Hyg 59:471–477

Gros P, Skamene E, Forget A (1981) Genetic control of natural resistance to *Mycobacterium bovis* (BCG) in mice. J Immunol 127:2417–2421

Gros P, Skamene E, Forget A (1983) Cellular mechanisms of genetically controlled host resistance to *Mycobacterium bovis* (BCG). J Immunol 131:1966–1972

Kakinuma M, Onoé K, Okada M, Kimura T, Kato K, Okuyama H, Morikawa K, Yamamoto K (1981) Failure of C3H mice to develop lung granuloma after intravenous injection of BCG cell wall vaccine. Demonstration of a defect in lymphoid cells. Immunology 43:1–9

Kakinuma M, Onoé K, Yasumizu R, Yamamoto K (1983) Strain differences in lung granuloma formation in response to a BCG cell-wall vaccine in mice. Failure of antigen presentation by low-responder macrophages. Immunology 50:423–431

Kaledin VI, Kurunov YN, Serova IA (1977) Inhibition and stimulation of the growth of Krebs-2 carcinoma by BCG vaccine. JNCI 58:1271–1277

Kaledin VI, Kurunov YN, Matienko NA, Nikolin VP (1978) Stimulation of tumor growth in mice by high doses of BCG. JNCI 61:1393–1396

Kato K, Yamamoto K (1982) Suppression of BCG cell wall induced delayed type hypersensitivity by BCG pretreatment. II. Induction of suppressor T cells by heat-killed BCG injection. Immunology 45:655

Lagrange PH, Hurtrel B (1979) The influence of BCG vaccination on murine leprosy in C57BL/6 and C3H mice. Ann Immunol (Paris) 130C:687

Lagrange PH, Hurtrel B, Thickstun PM (1979) Immunological behavior after mycobacterial infection in selected lines of mice with high or low antibody responses. Infect Immun 25:39–47

Lagrange PH, Hurtrel B, Brandely M, Thickstun PM (1983) Immunological mechanisms controlling mycobacterial infections. Bull Eur Physiopathol Respir 19:163–172

Lefford MJ (1970) The effect of inoculum size on the immune response to BCG infection in mice. Immunology 21:369–381

Lissner CR, Swanson RN, O'Brien AD (1983) Genetic control of the innate resistance of mice to *Salmonella typhimurium*. Expression of the *Ity* gene in peritoneal and splenic macrophages isolated in vitro. J Immunol 131:3006–3013

Lurie MB, Dannenberg AM (1965) Macrophage function in infectious disease with inbred rabbits. Bacter Rev 29:466–476

Lurie MB, Abramson S, Heppleston AG (1952a) On the response of genetically resistant and susceptible rabbits to the quantitative inhalation of human-type tubercle bacilli and the nature of resistance to tuberculosis. J Exp Med 95:119–134

Lurie MB, Zappasodi P, Dannenberg AM Jr, Weiss GH (1952b) On the mechanism of genetic resistance to tuberculosis and its mode of inheritance. Am J Hum Genet 4:302–314

Lurie MB, Zappasodi P, Cardona-Lynch E, Dannenberg AM Jr (1952c) The response to the intracutaneous inoculation of BCG as an index of native resistance to tuberculosis. J Immunol 68:369–387

Lynch CJ, Pierce-Chase CH, Dubos R (1965) A genetic study of susceptibility to experimental tuberculosis in mice infected with mammalian tubercle bacilli. J Exp Med 121:1051–1070

Medical Research Council (1963) BCG and role bacillus vaccines in the prevention of tuberculosis in adolescence and early adult life. Br Med J I:973

Moore VL, Mondloch VM, Pedersen GM, Schrier DL, Allen EM (1981) Strain variation in BCG-induced chronic pulmonary inflammation in mice: control by cyclophosphamide-sensitive thymus derived suppressor cells. J Immunol 127:339–342

Motulsky AG (1960) Metabolic polymorphisms and the role of infectious disease in human evolution. Hum Biol 32:28–62

Nakamura RM, Tokunaga T (1978) Strain difference of delayed-type hypersensitivity to BCG and its genetic control in mice. Infect Immun 22:657–664

Nakamura RM, Tanaka H, Tokunaga T (1982a) Strain difference in delayed-type hypersensitivity to BCG in mice: role of splenic adherent cells in the primary immune response. Immunology 47:729–731

Nakamura RM, Tanaka H, Tokunaga T (1982b) In vitro induction of suppressor T-cells in delayed-type hypersensitivity to BCG and an essential role of I-J positive accessory cells. Immunol Lett 4:295–299

Neta R, Salvin SB (1980) In vivo release of lymphokines in different strains of mice. Cell Immunol 51:173–178

Pelletier M, Forget A, Bourassa D, Gros P, Skamene E (1982) Immunopathology of BCG infection in genetically resistant and susceptible mouse strains. J Immunol 129:2179–2185

Potter M, O'Brien AD, Skamene E, Gros P, Forget A, Kongshavn PAL, Wax J (1983) A BALB/c congenic strain of mice that carries a genetic locus (*Ity*^{r}) controlling resistance to intracellular parasites. Infect Immun 40:1234–1235

Rook GAW, Rainbow S (1981) An isotope incorporation assay for the antimycobacterial effects of human monocytes. Ann Immunol (Paris) 132D:281

Schrier DJ, Allen EM, Moore VL (1980) BCG-induced macrophage suppression in mice: suppression of specific and nonspecific antibody-mediated and cellular immunologic responses. Cell Immunol 56:347–352

Schrier DJ, Sternick JL, Allen EM, Moore VL (1982) Immunogenetics of BCG-induced anergy in mice. Control by genes linked to the IgH complex. J Immunol 128:1466–1470

Schweitzer MD (1961) Genetic determinants of communicable disease. Ann NY Acad Sci 91:730–757

Sever JL, Youmans GP (1957) The enumeration of non pathogenic viable tubercle bacilli from the organs of mice. Am Rev Tubercol 75:280–294

Skamene E (1983) Genetic regulation of host resistance to bacterial infection. Rev Infect Dis 5:S823–S832

Skamene E, Gros P, Forget A, Kongshavn PAL, St Charles C, Taylor BA (1982) Genetic regulation of resistance to intracellular pathogens. Nature 297:506–510

Skamene E, Gros P, Forget A, Patel PJ, Nesbitt MN (1984) Regulation of resistance to leprosy by chromosome 1 locus in the mouse. Immunogenetics 19:117–124

Stach J-L, Gros P, Forget A, Skamene E (1984) Phenotypic expression of genetically-controlled natural resistance to *Mycobacterium bovis* (BCG). J Immunol 132:888–892

Tokunaga T, Yamamoto S, Nakamura RM, Kurosawa A, Murohashi T (1978) Mouse-strain difference in immunoprophylactic and immunotherapeutic effects of BCG on carcinogen-induced autochthonous tumors. Jpn J Med Sci Biol 31:143–154

Wiener E, Dandieri A (1974) Differences in antigen handling by peritoneal macrophages from the Biozzi high and low responder mice. Eur J Immunol 4:457–463

Yamamoto K, Kakinuma M (1978) Genetic control of granuloma response to oil-associated BCG cell wall vaccine in mice. Microbiol Immunol 22:335–348

Youmans GP, Youmans AS (1972) Response of vaccinated and non-vaccinated syngeneic C57BL/6 mice to infection with *Mycobacterium tuberculosis*. Infect Immun 6:748–754

Genetic Control of Resistance to *Listeria* Infection

P.A.L. KONGSHAVN

1 Introduction

Listeriosis is a disease which occurs in animals and, infrequently, in humans (DUMONT and COTONI 1921; MURRAY et al. 1926). The causative agent, *Listeria monocytogenes*, is a facultative, intracellular bacterium that parasitizes host macrophages. It is a gram-positive coccobacillus that is widely distributed in nature, having been isolated from soil, water, and many species of mammals, fowl, and fish. In humans, bacterial meningitis is the most common form of listeriosis in adults, while perinatal infections may result in abortion, stillbirth, or infant death (SEELIGER 1981). A predisposition to developing *Listeria* infection is seen in immunocompromised hosts, such as renal transplant patients or those with malignancy (NIEMAN and LORBER 1980).

The very elegant studies of MACKANESS and coworkers in the 1960s and early 1970s established experimental infection of rodents with *Listeria* as a model

Department of Physiology, 3655 Drummond Street, Montreal, Quebec, H3G 1Y6, Canada

Current Topics in Microbiology and Immunology, Vol. 124
© Springer-Verlag Berlin·Heidelberg 1986

infection in which the acquisition of cellular resistance could be analyzed (MACK-ANESS 1962, 1969; McGREGOR et al. 1971). Numerous studies have been carried out by other investigators using this model and, in recent years, it has been employed to investigate genetic regulation of the steps involved in the development of cellular resistance to listeriosis. The present review will focus on this aspect of the studies.

2 Experimental Model of Listeriosis

2.1 Bacterial Growth Kinetics

When mice are exposed to a sublethal inoculum of *Listeria*, the ensuing infection follows a well-defined course, lasting for approximately 1 week. Intraperitoneal, oral, subcutaneous, or other routes of inoculation produce infection (AUDURIER et al. 1980) but the intravenous route is used most commonly. In a typical experiment, mice are injected intravenously with a sublethal dose of *Listeria* and the bacterial growth kinetics monitored subsequently in the spleen and liver (MACKANESS 1962; MITSUYAMA et al. 1978). It is observed that, within 10 min of injection, 90% of the inoculum is taken up by the liver and 5%–10% by the spleen. Six hours later, the number of viable *Listeria* in the liver has decreased ten fold, indicating rapid destruction of most of the bacteria. Surviving organisms, however, start to multiply within susceptible macrophages and growth in the spleen and liver is logarithmic for the next 48 h, peaking on day 2 or 3 postinfection (p.i.). Rapid bacterial inactivation ensues over the next 3–4 days, signaling recovery of the host. Convalescent mice are resistant to challenge infection.

2.2 Host Response: Natural and Acquired Resistance

Recovery from infection with *Listeria* is associated with the development of acquired cellular resistance (ACR), the process in which the soluble products (lymphokines) of T-lymphocytes specifically sensitized to *Listeria* antigens induce effector macrophages to express powerful microbicidal activity and to form granulomatous lesions at infective foci (MACKANESS 1969; LANE and UN-ANUE 1972; BLANDEN and LANGMAN 1972; NORTH 1973; McGREGOR and KOS-TIALA 1976). The rapid bacterial inactivation observed in the spleen and liver thus provides a measure of the expression of ACR. Production of the specific T-cell mediators of ACR in infected hosts can be quantified more precisely by assessing the protective capacity of lymphoid cells (from the spleen, peritoneal cavity, or thoracic duct) adoptively to immunize naive recipients against a lethal challenge infection (MACKANESS 1969; McGREGOR and KOSTIALA 1973; NORTH 1973). Peak production of these cells normally occurs on the 6th day of infection. The magnitude and kinetics of the delayed-type hypersensitivity (DTH) reaction to *Listeria* antigens, as measured by the 24-h footpad swelling, usually corresponds directly with the level of ACR (MACKANESS 1969; NORTH 1969), inferring that both phenomena are different manifestations of the same cell-mediated response to microbial antigens, mediated by the same population of T-lymphocytes. Certainly, MITSUYAMA (1982) recently demonstrated that local challenge with live *Listeria* at the site of a delayed footpad reaction resulted in bacterial destruction at the site, indicating that the cellular infiltrate associated with DTH has the capacity to express enhanced antimicrobial resistance. In contrast, several reports have shown the dissociated development of DTH and ACR, suggesting that these two parameters of specific immunity can occur as independent phenomena, mediated by different subpopulations of T-lymphocytes (OSEBOLD et al. 1974; KERCKHAERT et al. 1977; VAN DE MEER et al. 1979; JUNGI et al. 1982). Recent evidence tends to support this idea, as will be discussed further in a later section (Sect. 4.3). In mice which are unable to generate ACR, such as athymic nude hosts, a persistent and eventually fatal form of listeriosis is seen, underscoring the need for specific cell-mediated immunity for ultimate cure (EMMERLING et al. 1975; TAKEYA et al. 1977; NEWBORG and NORTH 1980; SCHAFFNER et al. 1983).

Humoral antibody makes no contribution to resistance since passive transfer of immune serum fails to protect the host against listerial infection (MIKI and MACKANESS 1964). Living *Listeria* organisms, rather than killed vaccines or bacterial extracts, are normally required to stimulate the development of acquired immunity (MACKANESS 1962; VON KOENIG et al. 1982) except in very special cases, e.g., the *Listeria* intracellular product described by KEARNS and DE FREITAS (1983).

There is no convincing evidence that specific immunity plays a protective role in the host during the early, logarithmic bacterial growth phase of the infection. Thus, DTH and the adoptive transfer of protection are detectable only between the 12 and 14 days (NORTH 1973; ZINKERNAGEL et al. 1974). A more compelling argument comes from a number of observations that, in mice which are unable to generate T-cell immunity, such as athymic nude hosts, bacterial growth over the first 2–3 days is not greater than in that of their normal counterparts (e.g., EMMERLING et al. 1975; TAKEYA et al. 1977; NEWBORG and NORTH 1980; SCHAFFNER et al. 1983). Yet, from indirect evidence, it can be deduced that the host *is* engendering a significant degree of protection against infection during this early period. Firstly, treatment of infected mice with macrophage-inactivating agents, such as dextran sulfate 500 (DS-500), silica, carbon particles, carrageenan, or X-irradiation, can lead to a drastic increase in early listerial growth, even in nude mice (TAKEYA et al. 1977; MITSUYAMA et al. 1978; SADARANGANI et al. 1980; NEWBORG and NORTH, 1980) and, conversely, stimulants of the mononuclear phagocyte system (MPS), for example, *Bordetella pertussis* organisms (FINGER et al. 1978) or bacille Calmette-Guérin (BCG) (MATSUO et al. 1981), lead to suppressed bacterial growth. Secondly, bacterial growth is significantly lower in some mouse strains that in others at this time, as will be discussed in more detail below (CHEERS et al. 1978; SKAMENE and KONGSHAVN 1979).

Therefore, protective immunity to infection with *Listeria* in the mouse can be viewed as comprising, first of all, a phase of natural or nonspecifically induced resistance, and this is followed after 2 or 3 days by an acquired or immunologically specific response. Both natural and acquired resistance are under genetic regulation.

3 Genetic Control of Natural Resistance to Listeriosis

3.1 Cellular Mediators of Natural Resistance

The cellular response of the host to listerial invasion during the phase of natural resistance has been analyzed with some precision by MITSUYAMA et al. (1978). Initial trapping and destruction of bacterial cells at the very early stage of infection (0–6 h p.i.) can be attributed to radioresistant, carrageenan-susceptible, resident macrophages (Kupffer cells), while suppression of fulminating bacterial growth during the subsequent 2- to 3-day period is ascribed to monocyte-derived, inflammatory macrophages, sensitive both to x-irradiation and carrageenan. The importance of these immigrant mononuclear phagocytes in providing effective antilisterial resistance had been demonstrated in earlier studies by NORTH (1970) in the mouse and by McGREGOR and LOGIE (1975) in the rat. Recent in vitro studies have confirmed the superior listericidal activity of the young mononuclear phagocyte, using human, as well as mouse, cells (CZUPRYNSKI et al. 1983, 1984).

The participation of neutrophilic polymorphonuclear leukocytes (PMNs) in antilisterial resistance has not been studied adequately owing to the lack of satisfactory methods for depleting animals specifically of this cell type. PMNs are always a prominent feature in focal lesions early in infection (MACKANESS 1962). However, bacterial multiplication occurs rapidly at this time and, in fact, is seen to take place largely in foci dominated by PMN, providing grounds for arguing that these cells make no contribution to antilisterial resistance (NORTH 1970). TATSUKAWA et al. (1979) also reported that the number of *Listeria* was not decreased in an intramuscular infective site containing 80% PMNs, whereas the PMN-sensitive organism *Pseudomonas aeruginosa* was markedly decreased in number, under similar conditions. And, in experiments in which the MPS of the host had been compromised by carrageenan, while leaving PMNs intact both numerically and functionally (as measured by antimicrobial activity toward *P. aeruginosa* or *Candida albicans*), susceptibility of such hosts to listerial infection was greatly increased during the phase of natural resistance (TATSUKAWA et al. 1979; HURTREL and LAGRANGE 1981). In contrast to these observations, experiments have shown that both human and mouse neutrophils readily kill *Listeria* in vitro, in fact, more effectively than mononuclear phagocytes under the same experimental conditions (MACGOWAN et al. 1983; CZUPRYNSKI et al. 1983; 1984). Therefore, until the role of PMNs in resistance to listeriosis can be disproved by direct experimentation, the contribution of these cells to host defense cannot be excluded.

A role for natural killer (NK) cells in providing natural antilisterial resistance seems unlikely from the conceptual point of view, since lysis of *Listeria*-infected target macrophages would only release bacteria ready to invade elsewhere. And, indeed, NK cell involvement in antilisterial resistance has not been observed. NK-deficient (beige) mice are as resistant to listerial infection as their normal counterparts and the strain distribution pattern (SDP) of susceptibility/resistance to *Listeria* does not correspond with that of high or low NK cell activity (SKAMENE and KONGSHAVN 1983; CHEERS and WOOD 1984).

In summary, a large body of evidence favors the MPS as the major mediator of natural antilisterial resistance, the fixed macrophage providing the initial bacterial destruction which occurs shortly following infection, followed by the monocyte-derived inflammatory macrophage which subsequently populates the foci of infection.

3.2 Genetic Basis of Susceptibility/Resistance

Inbred mouse strains had been noted to vary in their responsiveness to listerial infection (ROBSON and VAS 1972) and, a number of years ago, studies were initiated independently both in our laboratory and elsewhere to investigate the genetic basis and phenotypic expression of resistance/susceptibility to murine listeriosis. When screened for this characteristic, using the median lethal dose of infection (LD_{50}) as the parameter of resistance, mouse strains fell into two categories, resistant strains having an LD_{50} which was approximately 100-fold higher than that of susceptible strains (CHEERS and McKENZIE 1978; CHEERS et al. 1978; SKAMENE and KONGSHAVN 1979; SKAMENE et al. 1982). Using a single typing dose ($6-12 \times 10^4$ organisms), the strains were classified into susceptible or resistant, based on the mean time of death: susceptible, 50% or fewer mice surviving day 5; resistant, more than 50% surviving day 14 (CHEERS and McKENZIE 1978) (Table 1).

Genetic analysis by CHEERS et al. (CHEERS and McKENZIE 1978; CHEERS et al. 1980) using backcross progeny of resistant C57BL and susceptible Balb/c strain mice, revealed that a single gene, named *Lr* (for *Listeria* resistance), was the major determinant of resistance/susceptibility to listeriosis. The (C57BL × Balb/c)F_1 hybrid was relatively resistant and, in all strains tested, there was no difference between males and females. Using backcross, congenic, and recombinant mice, linkage of *Lr* to a variety of known genetic markers, namely, *H-1*, *H-2*, *H-3*, *H-4*, *H-7*, and *H-8* loci, *Ig-1* allotype, *Thy-1* gene, the *Hc* gene specifying C5, and coat color genes (*B*, *c*) was excluded. Gene complementation studies indicated that susceptibility to *Listeria* in the CBA strain was under the same genetic control. The possibility that other genes exerted a modifying influence on resistance/susceptibility to *Listeria* infection was not excluded.

Concurrent studies in our laboratory, selecting A/J as the susceptible prototype, revealed a similar picture in which resistance/susceptibility to listeriosis was governed by a major, autosomal, dominant gene (SKAMENE and KONGSHAVN 1979; KONGSHAVN et al. 1980a). Very recently, the trait of *Listeria* resistance/susceptibility has been typed in AXB/BXA recombinant inbred (RI) strains derived from A/J (A) and C57BL/6 (B) progenitors (GERVAIS et al. 1984) and

Table 1. Strain survey for resistance and susceptibility to *Listeria*[a]. (CHEERS and McKENZIE 1978)

Strain	% Mice surviving		Median time to death[b] (days)
	5 days	14 days	
Resistant			
C57BL/6J	100	100	
C57BL/10ScSn	100	100	
B10.D2/Sn	100	60	
B10.A/SnSg	100	60	
(B10.A × A)F$_1$	100	100	
NZB/WEHI	100	80	
B6.C-*H-2*^{d} By	100	88	
SJL/WEHI[c]	100	100	
B6.PL (74NS)/Cy	100	90	
Susceptible			
Balb/cJ	0	0	3
CBA/H	0	0	3
DBA/IJ	0	0	3
C3H.OH/Sf	0	0	3
C3H/HeJ	0	0	4
C3H.OL/Sf	25	13	4
A/J[c]	0	0	4
LP.RIII	40	40	5
WB/Re	50	13	5
129/J	33	17	5

[a] Doses of *Listeria* varied between 6×10^4 and 12×10^4 in different experiments. Between five and ten mice were used per group. C57BL/6 and Balb/c controls were included in each experiment
[b] Time taken for 50% of mice to die was noted for susceptible mice. However, by definition, less than 50% of resistant mice died
[c] Only male SJL and A/J mice were available. All others shown were female

the SDP of this trait has essentially confirmed the earlier observation. The results suggest that an allelic difference at a major locus, now (since 1984) termed *Lr-1*, is responsible for listerial resistance/susceptibility. In addition, however, a putative minor gene (*Lr-2*), unlinked to *Lr-1*, is postulated also to control the level of the bacterial load within the susceptible strains (Fig. 1). Thus, the presence at the *Lr-1* locus of an A-derived allele (from A/J) would render the type of response that is classified as susceptible, i.e., a bacterial burden significantly above that seen in the C56BL/6 progenitor carrying the B allele. The degree of susceptibility (among the group of susceptible strains) is then regulated by the presence of either A- or B-derived alleles at the putative *Lr-2* locus: those strains carrying the A allele are classified as fully susceptible, whereas the presence of the B allele renders the strain semisusceptible. As will be discussed later, the major *Listeria* resistance genes controlling susceptibility of Balb/c (*Lr*) and of A/J (*Lr-1*) strains of mice would appear to be different.

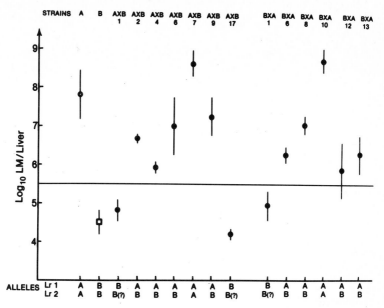

Fig. 1. Strain distribution pattern of resistance/susceptibility to *Listeria* in AXB/BXA recombinant inbred strains derived from A/J (*A*) and C57BL/6 (*B*) progenitors. The bacterial burden of *Listeria* in the livers is determined 72 h after inoculation with $1-2.5 \times 10^4$ CFUs *Listeria*. The results are expressed as the mean $\log_{10}$ CFUs *Listeria* ± SEM. *Horizontal line*, 99% limit of confidence for typing the strain as resistant (*below line*) or susceptible (*above line*). *A* (from A/J) and *B* (from C57BL/6) alleles for the major gene (*Lr-1*) and the putative minor gene (*Lr-2*) have been assigned as shown. For further details, see the original article. (GERVAIS et al. 1984)

It is clear from experiments measuring antilisterial resistance in *H-2* congenic and recombinant strains of mice that the *Listeria* resistance genes are not linked to the major histocompatibility complex (MHC) controlling specific immune responses (CHEERS and MCKENZIE 1978; SKAMENE et al. 1979).

3.3 Phenotypic Expression of *Listeria* Resistance Genes in Relation to Host Response

The phenotypic expression of the *Listeria* resistance genes was examined by comparing the course of infection in susceptible and resistant prototype mouse strains over a wide range of infective doses: either A/J and C57BL/6 (SKAMENE and KONGSHAVN 1979) or Balb/c and C57BL (CHEERS et al. 1978). In both studies, the initial uptake and rapid destruction of most of the injected inoculum by the liver was observed to occur equally well in resistant and susceptible hosts. By 24 h postinfection, however, the genetically determined susceptibility to *Listeria* became apparent in Balb/c or A/J strain mice and was clearly evident during the phase of logarithmic bacterial growth (Fig. 2); as the challenge dose of *Listeria* was increased, the strain difference in susceptibility became more

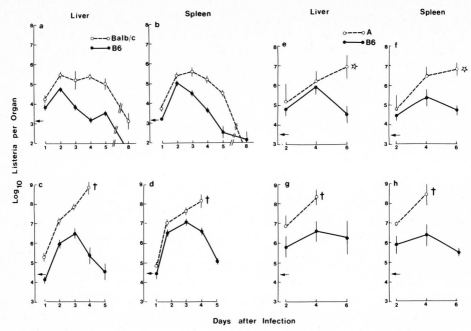

Fig. 2 a–h. Growth curves of *Listeria* in liver and spleen of genetically resistant C57BL/6 (B6) (*solid line*) and genetically susceptible Balb/c (*dotted line*) or A/J (*dotted line*) strain mice following challenge with various doses of *Listeria* (*arrows*) on day 0. **a–d** Mean of five mice ± 1 SEM. † signifies that no Balb/c mice survived beyond day 4. Redrawn from CHEERS et al. (1978). **e–h** Mean of six to eight mice ± 1 SEM. † denotes four of six mice died by day 4; * denotes four of eight mice died by day 6. Redrawn from SKAMENE and KONGSHAVN (1979)

marked. Thus, with a relatively high inoculum (4–5 $\log_{10}$ organisms), mice of the susceptible genotype failed to prevent bacterial numbers from reaching a lethal level (10^8–10^9) and they died (Fig. 2c, d, g, h). Mice of the resistant genotype, on the other hand, were able to hold bacterial multiplication in check until such time as ACR developed, around the 2nd or 3rd day, leading to recovery of the host.

These differences in early bacterial growth kinetics between the strains leave little doubt that the *Listeria* resistance genes are associated primarily with the expression of natural resistance to listeriosis: a function of the MPS. Confirmation that the resistance genes are expressed in the MPS, rather than in the T cell, was obtained by CHEERS et al. (1980) in T-cell chimera experiments showing that the T-cell genotype (resistant or susceptible) failed to influence the early bacterial growth kinetics and, in our laboratory, by the demonstration that *Listeria*-immune T cells (of either genotype) could adoptively transfer protection to naive recipients (by activating their macrophages) much more efficiently when the latter were of the resistant, rather than the susceptible, genotype (KONGSHAVN et al. 1980a).

Susceptible mice have the potential to develop powerful specific immunity. For instance, the LD_{50} for the secondary response is greatly improved over that for the primary infection (CHEERS et al. 1978; SKAMENE and KONGSHAVN 1979). Also, adoptive transfer of potent ACR can be achieved by susceptible-type T cells under the right circumstances (CHEERS et al. 1978; KONGSHAVN et al. 1980a), for instance, if the bacterial counts in the recipient are measured on day 3 rather than on day 2 p.i. (CHEERS et al. 1978).

3.4 Phenotypic Expression of *Listeria* Resistance Genes at the Cellular Level

3.4.1 Susceptibility of the A/J Strain

The precise manner in which the genetic control of antilisterial resistance is being expressed in relation to the MPS is, naturally, a key issue. As already indicated, two mononuclear cells contribute to natural antilisterial resistance. One is the radioresistant, fixed macrophage, responsible for the bacterial decimation seen shortly (0–6 h) following inoculation. Since this process occurs equally well in resistant and susceptible mice, the function of the fixed macrophage appears not to be influenced by the *Listeria* resistance genes. The second cell is the monocyte-derived, inflammatory macrophage which populates infective foci; the functional activity of this cell is known to be highly radiosensitive (VOLKMAN and COLLINS 1971). There is no large preformed store of monocytes so that an influx of inflammatory macrophages into foci of infection must be preceded by a wave of mitosis in the bone marrow progenitor compartment, and this process is generally interpreted as being the radiation-sensitive target. Experimental evidence indicated that the high natural resistance of C57BL/6 mice to *Listeria* could be attributed to a prompt influx of inflammatory macrophages which were able to control the early, rapid, bacterial multiplication in infective sites, since whole body irradiation, with a dose as low as 200 R, totally eliminated the genetic advantage of the resistant animals at this time, and transfer of syngeneic bone marrow cells fully restored this property (SADAR-ANGANI et al. 1980). The high natural resistance of another strain, SJL, was similarly found to be radiosensitive (KONGSHAVN et al. 1980a). In contrast, irradiation had no effect on natural resistance in A/J strain mice, indicating that their susceptibility to listerial infection was due to a paucity of inflammatory macrophages accumulating in infective foci at this time.

This hypothesis was strengthened by showing that the MPS responded promptly to listerial infection in the resistant host during the phase of natural resistance while failing to do so in the susceptible A/J hosts. These data are summarized in Table 2: mononuclear phagocyte production, emigration, and accumulation at infective foci are all increased in the resistant host shortly following infection (PUNJABI et al. 1984; KONGSHAVN and SKAMENE 1984).

The implication that an efficient macrophage inflammatory response is linked to listerial resistance was formally proved to be correct by backcross analysis in progeny of resistant B10.A and susceptible A/J mouse strains (STE-VENSON et al. 1981). A strain survey indicated that the magnitude of the macrophage inflammatory response, as quantified by the number of macrophages

Table 2. Early response of mononuclear phagocyte system (MPS) to *Listeria* infection in resistant and susceptible mice. Data from PUNJABI et al. (1984) and KONGSHAVN (unpublished)

MPS compart-ment	Parameter measured	Resistant mice		Susceptible mice		Effect seen in resistant mice
		Control	Infected	Control	Infected	
Bone marrow	T_G(h)	17.3	13.7	18.0	20.0	Increased production of mononuclear phagocytes
Blood	Mono-cytes/mm^3 24 h 48 h	200 ± 64 198 ± 72	834 ± 98 710 ± 75	140 ± 77 280 ± 88	278 ± 62 182 ± 61	Increased number of monocytes in blood
Blood	$T_{1/2}$(h)	30.0	21.1	32.0	30.4	Rapid emigration of young macrophages
Tissues	$M\phi \times 10^{-3}$ 24 h 72 h	0 0	520 ± 90 960 ± 200	0 0	50 ± 10 90 ± 10	Accumulation of in-flammatory macrophages in infective foci

Resistant mouse strains: C57BL/6 or B10.A. Susceptible mouse strain: A/J. T_G, promonocyte generation time 8 h p.i. $T_{1/2}$, half-time of blood monocytes measured starting 24 h p.i. $M\phi$, # macrophages in s.c. site of listerial infection, induced in a pocket prepared in the dorsum of a mouse, measured at 24 h and 72 h p.i. Doses of *Listeria* were 1×10^4 to 5×10^4 colony-forming units (CFUs) for T_G, monocyte, and $T_{1/2}$ values. For mϕ/infected site, 1.25×10^6 CFUs *Listeria* were injected into the pocket

found in the peritoneal cavity in response to injection of a phlogistic agent such as thioglycolate, was controlled by an autosomal, dominant gene mapping outside the MHC. This gene was linked, or identical, to the major gene determining *Listeria* susceptibility in the A/J mouse strain. Genetic linkage between (1) resistance/susceptibility to *Listeria* and (2) macrophage inflammatory responsiveness, was confirmed by showing cosegregation of both traits in AXB/BXA RI strains (GERVAIS et al. 1984; STEVENSON et al. 1984).

An allelic difference in the *Hc* locus, which specifies the C5 component of complement, results in deficiency in C5 levels in the A/J (*Hc*^{0}) strain of mice while C57BL/6 (*Hc*^{1}) strain mice are C5 sufficient (CINADER et al. 1964). C5a is, of course, recognized as an important chemoattractant for macrophages and PMNs during inflammation. When the SDP of the *Hc* locus in the RI strains was measured by the presence/absence of hemolytic serum activity, it was found to be fully concordant with the other traits under consideration, i.e., all strains which typed as susceptible to *Listeria* and had low macrophage inflammatory responses, were deficient in C5 (GERVAIS et al. 1984; STEVENSON et al. 1984). These data are shown in Table 3. The PMN inflammatory response was also found to be under the same genetic regulation as that for the macrophage but, as already discussed, its role in antilisterial resistance has not been established.

It can be concluded, therefore, that a defect in the phagocyte inflammatory response, caused by C5 deficiency, is the major reason for the extreme susceptibility of A/J mice to listeriosis (GERVAIS et al. 1984); in other words, *Lr-1* is the *Hc* locus. DBA/2 (*Hc*^{0}) strain mice (CINADER et al. 1964) could presumably be susceptible to *Listeria* infection for the same reason.

Table 3. Resistance to *Listeria monocytogenes*, macrophage inflammatory responses, and complement levels in AXB/BXA RI strains. (STEVENSON et al. 1984)

Strain	Number of CFUs/liver[a] (mean ± SEM)	Inflammatory response[b] (mean macrophages × 10^6/ mouse ± SEM)	Complement[c] (% mean lysis ± SEM)
C57BL/6J	4.5 ± 0.25	21.6 ± 1.11	99.7 ± 2.2
A/J	7.8 ± 0.60	6.7 ± 0.51	4.0 ± 0.81
AXB-1	4.8 ± 0.25	16.8 ± 1.01	89.5 ± 3.5
AXB-2	6.67 ± 0.14	12.2 ± 1.08	2.3 ± 0.7
AXB-4	5.93 ± 0.16	8.8 ± 0.06	
AXB-6	6.99 ± 0.76	9.5 ± 0.8	7.0 ± 1.0
AXB-7	8.61 ± 0.3	10.8 ± 1.21	2.0 ± 1.0
AXB-9	7.19 ± 0.46	11.6 ± 2.1	3.5 ± 2.5
AXB-17	4.19 ± 0.09	15.2 ± 1.7	96.5 ± 3.5
BXA-1	4.95 ± 0.32	21.3 ± 2.2	94.0 ± 1.0
BXA-6	6.25 ± 0.17	6.5 ± 0.7	7.5 ± 1.5
BXA-8	7.03 ± 0.2	13.7 ± 1.8	6.0 ± 3.0
BXA-10	8.70 ± 0.3		4.0
BXA-12	5.86 ± 0.7	13.7 ± 1.2	3.0 ± 1.0
BXA-13	6.25 ± 0.5	8.8 ± 1.6	3.5 ± 0.5

[a] Resistant, $\leq \log_{10}$ 5.5 CFUs; 8–30 animals per strain.
[b] High level of macrophage accumulation, $\geq 15.0 \times 10^6$; 5–25 animals per strain.
[c] Complement sufficient, $\geq 90\%$ lysis; three to five mice per determination

High or low antilisterial resistance appears to be expressed as a property of the host environment rather than of the macrophage per se, according to an earlier observation (KONGSHAVN et al. 1980b). This was demonstrated in radiation bone marrow chimeras in which *Listeria*-resistant B10.A strain hosts, repopulated with marrow from *Listeria*-susceptible A/J donors, were resistant to listeriosis and vice versa (KONGSHAVN et al. 1980b). Since macrophages are the source of C5, it is hard to interpret this observation in the light of the recent findings until further experiments are carried out.

It appears from previous studies employing the *Hc* congenic lines (H^0C5−) B10.D2/oSn and (Hc^1C5+)B10.D2/nSn that C5 deficiency has only a marginal effect on *Listeria* resistance when it occurs on the genetic background of the C57BL-derived strains (LAWRENCE and SCHELL 1978; PETIT 1980); also CHEERS and MCKENZIE (1978) failed to demonstrate linkage between *Lr* and *Hc*, using the same congenic pair. However, these strains carry the effective putative *Lr-2* allele, which can correct to some extent the susceptibility to listeriosis induced by deficiency in C5 (see above). It remains to be established whether the presence of a susceptible allele at the putative *Lr-2* locus could overcome the host defense function of C5 in listeriosis and thus be responsible for the susceptibility of some C5+ strains (such as Balb/c) to this infection, or whether yet another locus (distinct from *Lr-2*) contributes to the overall control of resistance to *Listeria* infection, as seems more likely (GERVAIS et al. 1984). Gene complementation studies should answer this question.

3.4.2 Susceptibility of the Balb/c Strain

In the studies of CHEERS and MACGEORGE (1982), the early listerial growth in Balb/c, like A/J, strain mice was unaffected by radiation, in contrast to the radiosensitivity of natural resistance in C57BL strain mice. The latter has been attributed by these authors to its deleterious effect on emigration or bactericidal activity of macrophages, rather than to its antiproliferative effect, since the antimitotic drug, vinblastine, failed to affect natural resistance in C57BL (or Balb/c) strain mice. Histopathological differences were seen between Balb/c and C57BL strains during the phase of natural resistance: susceptible mice had many infective foci in the liver surrounded by few PMNs and a PMN leukocytosis was observed, whereas resistant mice had few foci surrounded by many PMNs and a monocytic response in the blood (MANDEL and CHEERS 1980). Without attempting a detailed interpretation of these findings, the fundamental point to be made is a simple one: susceptibility to listeriosis in Balb/c, as in A/J, strain mice seems attributable to a sluggish macrophage response to infection, whatever the reason.

3.4.3 Genetic Control of Macrophage Response to Infection:
Expression of Listeria Resistance Genes

Since resistance to listerial infection appears to be causally related to an early influx of inflammatory macrophages into foci of infection, and vice versa, any defective step in the complex sequence of events leading to the provision of such macrophages in focal lesions could theoretically manifest itself as a *Listeria* resistance gene. Genetically controlled differences between various mouse strains are already known to exist for a number of these steps: the proliferative kinetics of the multipotential stem cell and the promonocyte, the provocation of a peripheral blood monocytosis, macrophage chemotaxis, and macrophage activation by a variety of stimuli (for review, see SKAMENE and STEVENSON 1985). Some of these points have already been discussed. Of particular pertinence, two independent laboratories have shown that the enhancement of monocytopoiesis by a saline-soluble extract of *Listeria* (SHUM and GALSWORTHY 1979) is under genetic regulation. Firstly, FIM (factor-inducing monocytopoiesis) is produced by mature macrophages in response to saline extract of *Listeria* (or to latex particles) and both C57BL (resistant) and CBA (susceptible) strain mice developed similar peak serum levels of this factor (SLUITER et al. 1984). However, CBA strain mice were unable to respond with increased monocytopoiesis to sera containing FIM whereas, in C57BL strain mice, these sera evoked a monocytosis. Thus, the genetically controlled inability of bone marrow monocytic precursors to respond to FIM by increased proliferation could be one of the mechanisms underlying genetic susceptibility to *Listeria* infection in the CBA (and possibly Balb/c) mouse strain. Secondly, EF (endogenous factor) is detected in the serum in response to injection of saline extract of *Listeria* in B10.A (resistant) strain mice; this factor induces proliferation of bone marrow monocyte precursors and a peripheral blood monocytosis (FEWSTER and GALS-

WORTHY 1986). A/J (susceptible) strain mice neither produce nor respond to EF. Conceivably, genetic control of EF production could be the expression of the putative *Lr-2* gene (see above) or yet another – as yet unidentified – *Listeria* resistance gene.

However, until genetic linkage analysis is performed, putative associations between such genetically controlled steps in the macrophage response to listerial infection and natural resistance to listeriosis remain speculative.

3.4.4 Cellular Expression of Listeria Resistance Genes in MPS

Expression of high or low natural resistance to listeriosis is probably associated with both quantitative and qualitative differences in the effector macrophage population in listerial focal lesions. Thus, qualitatively, the newly emigrated mononuclear phagocyte is strongly implicated as the superior listericidal cell (see Sect. 3.1); resident peritoneal macrophages, by comparison, demonstrate inferior listericidal activity in vitro (GODFREY et al. 1983; CZUPRYNSKI et al. 1984). Solid evidence for an interstrain intrinsic difference in listericidal activity of mononuclear phagocytes is lacking. This makes an interesting contrast to the expression in macrophages of some of the other "resistance" genes covered in this volume, e.g., the *Ity* locus controlling natural resistance to *Salmonella, Leishmania*, and BCG, which is expressed as the intrinsic microbicidal property of the resident macrophage. Presumably, the quantity of macrophages recruited also affects the degree of antilisterial resistance.

Thus, the genes controlling *Listeria* resistance/susceptibility are suggested to be those that, by a variety of ways, control MPS function and thereby regulate recruitment of immigrant macrophages [and possibly PMNs also (CZU-PRYNSKI et al. 1984)] into focal lesions.

4 Genetic Control of ACR to Listeriosis

4.1 Cellular Mediators of ACR

Earlier work has shown that acquired immunity to listeriosis is mediated by a population of specifically sensitized, short-lived, nonrecirculating T cells which have a tendency to localize in inflammatory foci (KOSTER et al. 1971; NORTH and SPITALNY 1974), where they serve to enhance the microbicidal activity of macrophages in sites of infection by release of MAF (macrophage-activating factor), MIF (migration inhibition factor), CF (chemotactic factor), and other soluble mediators (reviewed in HAHN and KAUFMANN 1981). Recent studies by HAHN, KAUFMANN, and coworkers (KAUFMANN et al. 1979; KAUFMANN and HAHN 1982; NÄHER et al. 1984), investigating T-cell subsets, led to the conclusion that only cells of the Lyt 123 + phenotype could readily transfer both adoptive protection and DTH systemically. Cloned Lyt 1 + cells could transfer systemic protection poorly, i.e., if given in very large numbers. However, cells of this phenotype readily demonstrated effector function in vitro, inducing inter-leukin-1 release and secreting lymphokines, and they also provided protection when transferred *locally* ino the infected site. An appealing interpretation of these findings (NÄHER et al. 1984) was that immature T cells of the Lyt 123+ phenotype possessed the aforementioned property to enter focal lesions, where they matured into nonmigratory Lyt 1+ cells, which were the actual effector T cells of antibacterial protection. Newer findings by these authors, however, have modified this view. They present evidence that two distinct T-cell subsets provide antilisterial protection: a MAF/MIF-secreting Lyt 1+ cell which acti-vates macrophages, functions in in vitro assays, but transfers systemic protection rather indifferently; and an Lyt 2+ (Lyt 1+2+) cell which transfer potent protection in the host by promoting granuloma fromation at infective foci (SPER-LING et al. 1984; NÄHER et al. 1985). These studies underscore the importance of granuloma formation, as compared with the mere provision of activated macrophages, in bringing about effective ACR.

4.2 Genetic Control of Specific Immunity to Listeriosis

Genetically determined strain differences in specific cell-mediated immunity to listerial infection have been little documented. There is some evidence that specif-ic immunity, as measured by footpad reactivity, is linked to the *H-2* haplotype, being higher in mice with $H-2^b$ than with $H-2^a$ or $H-2^d$ haplotypes (SKAMENE et al. 1979). In studies with Balb/c mice, the onset of ACR was delayed by 24–48 h when compared with C57BL mice, as measured by the usual parameters (outlined in Sect. 2.2): the fall in bacterial numbers, the adoptive transfer of immunity, and the DTH footpad reactivity (CHEERS et al. 1978). This delayed onset of ACR seems most likely to be related to the expression of *Lr* in the macrophage component, i.e., the slow tempo of the MPS response to the T-cell lymphokines. Alternatively, this phenomenon could be related to the differences in *H-2* haplotypes between Balb/c ($H-2^d$) and C57BL ($H-2^b$) and thus be a manifestation of MHC-linked genetic control of specific immunity to listeriosis.

4.3 Major Histocompatibility Complex Restriction
of Specific CMI to *Listeria*

Genetic restriction of specific CMI to listeria was first linked to the I region of the MHC. Several years ago, ZINKERNAGEL et al. (1977) demonstrated that, in the adoptive transfer of protection to listerial infection, cell donors and recipi-ents needed to share *H-2I* identity. FARR et al. subsequently showed that the in vitro response of immune T cells to heat-killed *Listeria*, as measured by T-cell proliferation and the release of lymphokines and monokines, was strictly dependent upon presentation of the bacteria by macrophages bearing the I region gene products – the Ia antigens. This process was I-A restricted. So, also, was the functional activity of the Lyt1+ cloned T-cell described above (Sect. 4.1) when tested both in vivo and in vitro (KAUFMANN and HAHN 1982).

Very elegant studies by UNANUE and associates have shown that listerial antigen is endocytosed by Ia+ macrophages, processed, recycled to the cell surface, and presented in association with Ia to the T-cell, the process apparently involving an amplification circuit in which Ia$^+$ macrophages are recruited by gamma-interferon released from the stimulated T-cells (reviewed by UNANUE 1984). Thus, the I-A region of the MHC is clearly important in controlling the expression of specific CMI to *Listeria* in the mouse.

CHEERS and SANDRIN (1983) have recently reexamined MHC restriction of the transfer of protective immunity to listeriosis and have made the intriguing observation that *H-2K*, rather than *I-A*, provides the restriction element. Several explanations are offered for this contradictory finding, in particular that the mouse strains used by these authors (unlike those used by ZINKERNAGEL et al. 1977) all had identical non-*H-2* background genes, since the latter themselves can restrict adoptive transfer of antilisterial protection (see below). CHEERS and SANDRIN suggest that, with regard to the in vitro findings, by using killed *Listeria* an *I-A*-restricted T-cell subset (Lyt 1 +) is being stimulated while, in the in vivo adoptive transfer of resistance, an Lyt 2 + *H-2K* restricted population of T-cells responding to live organisms is being triggered. The newly described Lyt 2 + T-cell subset of Hahn and associates (Sect. 4.1) lends strong credence to this idea since this cell, which is able to transfer a high degree of protection associated with granuloma formation, shows genetic restriction at the *H-2K* region and apparently responds only to antigens from living bacteria (NÄHER et al. 1985). Further support for two MHC-restricted events comes from the work of JUNGI et al. (1982) using the rat model. The major restriction element in the response to killed *Listeria* antigens was the B region of *RT1* (equivalent to mouse I region) whereas the A region (equivalent to mouse *H-2K*) and, to a lesser extent, the B region, mediated adoptive transfer of antilisterial resistance and DTH to live organisms (summarized in Table 7 in JUNGI et al. 1982). This could explain (1) the frequent observation that dead *Listeria* is unable to evoke protection against challenge with live organisms (i.e., the poorly protective T-cell subset would have been generated) and (2) the occasional reports showing dissociated development of ACR and DTH (Sect. 2.2).

In summary, ACR seems to depend upon both class I- and class II-restricted T-cell subsets: an *H-2K*-restricted Lyt 2 + cell recognizing products of living *Listeria* and an *I-A*-restricted Lyt 1 + cell that responds to heat-killed *Listeria*. The functional distinction between these subsets has been discussed.

4.4 Non-MHC Restriction of Specific CMI to *Listeria*

Passive transfer of both DTH (BERCHE and NORTH 1982) and protective immunity (CHEERS and McKENZIE 1983) to *Listeria* can be restricted, not only by incompatibility at the *H-2* locus, but also by non-*H-2* genes. These data were obtained in transfer studies between *H-2* compatible strains with different background genes and also between parental strains and their F_1 hybrids. In the transfer of protection, antigens from *H-4* and *H-8* loci were identified as producing strong restriction. The mechanism responsible for this allogeneic restriction has not been explained.

5   Conclusions

Resistance of mice to listeriosis is controlled by genes which are expressed in the macrophage response to this infection. Thus, they become evident in the phase of natural resistance. Efficient T-cell macrophage cooperation in ACR is genetically controlled in that it is restricted by the MHC and also, unexpectedly, by the presence of non-*H-2* loci. In the long term, ACR is necessary for cure in listeriosis. However, it takes several days to develop and the host may die of a fulminating infection in the meantime. This is particularly true in relation to bacteria with short generation times, such as *Listeria*. Therefore, the expression of the *Listeria* resistance genes during natural resistance can be an equally decisive factor in determining the outcome of the infection. Extrapolation of these findings to human listeriosis remains speculative at the present time.

References

Audurier A, Pardon P, Marly J, Lantier F (1980) Experimental infection of mice with *Listeria monocytogenes* and *L. innocua*. Ann Microbiol (Paris) 131 B:47–57

Berche PA, North RJ (1982) Non-H-2 restriction of expression of passively transferred delayed sensitivity. J Exp Med 155:1334–1343

Blanden RV, Langman RE (1972) Cell-mediated immunity to bacterial infections in the mouse. Thymus-derived cells as effectors of acquired resistance to *Listeria monocytogenes*. Scand J Immunol 1:379–391

Cheers C, MacGeorge J (1982) Genetic and cellular mechanisms of natural resistance to intracellular bacteria. In: NK cells and other natural effector cells. Ed. Herberman R, Academic, New York

Cheers C, McKenzie IFC (1978) Resistance and susceptibility of mice to bacterial infection: genetics of listeriosis. Infect Immun 19:755–762

Cheers C, McKenzie IFC (1983) Restriction in adoptive transfer of resistance to *Listeria monocytogenes*. I. Influence of non-H-2 loci. Cell Immunol 76:304–310

Cheers C, Sandrin MS (1983) Restriction in adoptive transfer of resistance to *Listeria monocytogenes*. II. Use of congenic and mutant mice show transfer to be *H-2K* restricted. Cell Immunol 78:199–205

Cheers C, Wood P (1984) Listeriosis in beige mice and their heterozygous littermates. Immunology 54:711–717

Cheers C, McKenzie IFC, Pavlov H, Waid C, York J (1978) Resistance and susceptibility of mice to bacterial infection: course of listeriosis in resistant or susceptible mice. Infect Immun 19:763–770

Cheers C, McKenzie IFC, Mandel TE, Chan YY (1980) A single gene (*Lr*) controlling natural resistance to murine listeriosis. In: Genetic control of natural resistance to infection and malignancy. Academic, New York

Cinader B, Dubliski S, Wardlaw AC (1964) Distribution, inheritance, and properties of an antigen, MuB1, and its relation to hemolytic complement. J Exp Med 120:897–924

Czupryuski CJ, Campbell PA, Henson PM (1983) Killing of *Listeria monocytogenes* by human neutrophils and monocytes, but not by monocyte-derived macrophages. J Reticuloendothel Soc 34:29–44

Czuprynski CJ, Henson PM, Campbell PA (1984) Killing of *Listeria monocytogenes* by inflammatory neutrophils and mononuclear phagocytes from immune and non-immune mice. J Leuk Biol 35:193–208

Dumont J, Cotoni L (1921) Bacille semblable à celui de rouget de porc rencontré dans le L.C.R. d'un méningitique. Ann Immunol (Paris) 35:625

Emmerling P, Finger H, Bockemühl J (1975) *Listeria monocytogenes* infection in nude mice. Infect Immun 12:437–439

Farr AG, Kiely J-M, Unanue ER (1979a) Macrophage-T cell interactions involving *Listeria monocytogenes* – role of the H-2 gene complex. J Immunol 122:2395–2404

Farr AG, Wechter WJ, Kiely J-M, Unanue ER (1979b) Induction of cytocidal macrophages after in vitro interactions between *Listeria*-immune T cells and macrophages – role of *H-2*. J Immunol 122:2405–2412

Fewster D, Galsworthy SB (1986) Humoral regulation of monocytopoiesis in mice with genetically determined resistance or sensitivity to listeriosis. (to be published)

Finger H, Heymer B, Wirsing C-H, Emmerling P, Hof H (1978) Reversion of dextran sulfate-induced loss of antibacterial resistance by *Bordetella pertussis*. Infect Immun 19:950–960

Gervais F, Stevenson MM, Skamene E (1984) Genetic control of resistance to *Listeria monocytogenes*: regulation of leukocyte inflammatory responses by the *Hc* locus. J Immunol 132:2078–2083

Godfrey RW, Horton PG, Wilder MS (1983) Time course of antilisterial activity by immunologically activated murine peritoneal macrophages. Infect Immun 39:532–539

Hahn H, Kaufmann SHE (1981) The role of cell-mediated immunity in bacterial infections. Rev Infect Dis 3:1221–1250

Hurtrel B, Lagrange PH (1981) Comparative effects of carrageenan on systemic candidiasis and listeriosis in mice. Clin Exp Immunol 44:355–358

Jungi TW, Gill TJ, Kunz HW, Jungi R (1982) Genetic control of cell-mediated immunity in the rat. III T cells restricted by the *RT1.A* locus recognize viable *Listeria* but not isolated bacterial antigens. J Immunogenetics 9:445–446

Kaufmann SHE, Hahn H (1982) Biological functions of T cell lines with specificity for the intracellular bacterium *Listeria monocytogenes* in vitro and in vivo. J Exp Med 155:1754–1765

Kaufmann S, Simon MS, Hahn H (1979) Specific Lyt 123 T cells are involved in protection against *Listeria monocytogenes* and in delayed-type hypersensitivity to listerial antigens. J Exp Med 150:1033–1038

Kearns RJ, deFreitas EC (1983) In vitro propagation of antigen-specific T lymphocytes that adoptively transfer resistance to *Listeria monocytogenes*. Infect Immun 40:713–719

Kerckhaert JAM, Hofhuls FMA, Willers JMN (1977) Influence of cyclophosphamide on delayed hypersensitivity and acquired cellular resistance to *Listeria monocytogenes* in the mouse. Immunology 32:1027–1032

Kongshavn PAL, Skamene ES (1984) The role of natural resistance in protection of the murine host from listeriosis. Clin Invest Med 7:253–257

Kongshavn PAL, Sadarangani C, Skamene E (1980a) Cellular mechanisms of genetically determined resistance to *Listeria monocytogenes*. In: Genetic control of natural resistance to infection and malignancy. Eds. Skamene E, Kongshavn PAL, Landy M. Academic, New York

Kongshavn PAL, Sadarangani C, Skamene E (1980b) Genetically determined differences in antibacterial activity of macrophages are expressed in the environment in which the macrophage precursors mature. Cell Immunol 53:341–349

Koster FT, McGregor DD, Mackaness GB (1971) The mediator of cellular immunity. II. Migration of immunologically committed lymphocytes into inflammatory exudates. J Exp Med 133:400–409

Lane FC, Unanue ER (1972) Requirement of thymus (T) lymphocytes for resistance to listeriosis. J Exp Med 135:1104–1112

Lawrence DA, Schell RF (1978) Susceptibility of C5-deficient mice to listeriosis: modulation by concanavalin A. Cell Immunol 39:336–344

MacGowan AP, Peterson PK, Keene W, Quie PG (1983) Human peritoneal macrophage phagocytic, killing, and chemiluminescent responses to opsonized *Listeria monocytogenes*. Infect Immun 40:440–443

Mackaness GB (1962) Cellular resistance to infection. J Exp Med 116:381–406

Mackaness GB (1969) The influence of immunologically committed lymphoid cells on macrophage activity in vivo. J Exp Med 129:973–992

Mandel TE, Cheers C (1980) Resistance and susceptibility of mice to bacterial infection: histopathology of listeriosis in resistant and susceptible strains. Infect Immun 30:851–861

Matsuo K, Takeya K, Nomoto K, Shimotori S, Terasaka R (1981) T-cell independent activation of macrophages by viable BCG in tumor-bearing mice. Cell Immunol 57:293–306

McGregor DD, Kostiala (1976) Role of lymphocytes in cellular resistance to infection. In: Contemporary topics in immunobiology. Volume 5. Ed. Weigle WO. Plenum, New York

McGregor DD, Logie PS (1975) The mediator of cellular immunity. X. Interaction of macrophages and specifically sensitized lymphocytes. Cell Immunol 18:454–465

McGregor DD, Koster FT, Mackaness GB (1971) The mediator of cellular immunity. I. The life-span and circulation dynamics of the immunologically committed lymphocyte. J Exp Med 133:389–399

Miki K, Mackaness GB (1964) The passive transfer of acquired resistance to *Listeria monocytogenes*. J Exp Med 120:93–104

Mitsuyama M, Takeya K, Nomoto K, Shimotori S (1978) Three phases of phagocyte contribution to resistance against *Listeria monocytogenes*. J Gen Microbiol 106:165–171

Mitsuyama M, Nomoto K, Takeya K (1982) Direct correlation between delayed footpad reaction and resistance to local bacterial infection. Infect Immun 36:72–79

Murray EGD, Webb RA, Swann MBR (1926) A disease of rabbits characterized by a large mononuclear leukocytosis caused by a hitherto undescribed bacillus: *Bacterium monocytogenes*. J Pathol Bacteriol 29:407–439

Näher H, Sperling U, Hahn H (1984) Developmental interrelationship of specific Lyt 123 and Lyt 1 cell sets in expression of antibacterial immunity to *Listeria monocytogenes*. Infect Immun 44:252–256

Näher H, Sperling U, Hahn H (1985) H-2K-restricted granuloma formation by Ly-2+ T cells in anti-bacterial protection to facultative intracellular bacteria. J Immunol 134:569–572

Newborg MF, North RJ (1980) On the mechanism of T cell-independent anti-*Listeria* resistance in nude mice. J Immunol 124:571–576

Nieman RE, Lorber B (1980) Listeriosis in adults: a changing pattern. Report of eight cases and review of the literature, 1968–1978. Rev Infect Dis 2:207–227

North RJ (1969) Cellular kinetics associated with the development of acquired cellular resistance. J Exp Med 130:299–314

North RJ (1970) The relative importance of blood monocytes and fixed macrophages to the expression of cell-mediated immunity to infection. J Exp Med 132:521–534

North RJ (1973) Cellular mediators of anti-*Listeria* immunity as an enlarged population of short-lived, replicating T cells. Kinetics of their production. J Exp Med 138:342–355

North RJ, Spitalny G (1974) Inflammatory lymphocytes in cell-mediated antibacterial immunity: factors governing the accumulation of mediator T cells in peritoneal exudates. Infect Immun 10:489–498

Osebold JW, Pearson LD, Medin NI (1974) Relationship of antimicrobial cellular immunity to delayed hypersensitivity in listeriosis. Infect Immun 9:354–362

Petit JC (1980) Resistance to listeriosis in mice that are deficient in the fifth component of complement. Infect Immun 27:61–67

Punjabi C, Galsworthy S, Kongshavn PAL (1984) Cytokinetics of mononuclear phagocyte response to listeriosis in genetically-determined sensitive and resistant murine hosts. Clin Invest Med 7:165–172

Robson HG, Vas SI (1972) Resistance of inbred mice to *Salmonella typhimurium*. J Infect Dis 126:378–386

Sadarangani C, Skamene E, Kongshavn PAL (1980) Cellular basis for genetically determined enhanced resistance of certain mouse strains to listeriosis. Infect Immun 28:381–386

Schaffner A, Douglas H, Davis CE (1983) Models of T cell deficiency in listeriosis: the effects of cortisone and cyclosporin A on normal and nude BALB/c mice. J Immunol 131:450–453

Seeliger HPR (1981) *Listeria monocytogenes*. In: Medical microbiology and infectious diseases, vol II. Eds. Brande AI, Davis CE, Fiever J. Saunders, Philadelphia

Shum DT, Galsworthy SB (1979) Stimulation of monocyte precursors in vivo by an extract from *Listeria monocytogenes*. Can J Microbiol 25:698–705

Skamene E, Kongshavn PAL (1979) Phenotypic expression of genetically controlled host resistance to *Listeria monocytogenes*. Infect Immun 25:345–351

Skamene E, Kongshavn PAL (1983) Cellular mechanisms of resistance to listeria. In: Host resistance to intracellular pathogens. Eds. Eisenstein T, Friedman H. Plenum, New York (Adv Exp Med 162)

Skamene E, Stevenson MM (1985) Genetic control of macrophage response to infection. In: Mononuclear phagocytes: characteristics, physiology and function. Ed. van Furth R. Nijhoss, Boston

Skamene E, Kongshavn PAL, Sachs DH (1979) Resistance to *Listeria monocytogenes* in mice: genetic control by genes that are not linked to the *H-2* complex. J Infect Dis 139:228–231

Skamene E, Stevenson MM, Kongshavn PAL (1982) Natural cell mediated immunity against bacteria. In: NK cells and other natural effector cells. Ed. Herberman R. Academic, New York

Sluiter W, Elzenga-Claasen I, van der Voort van der Kley-van Andel A, van Furth R (1984) Differences in the response of inbred mouse strains to the factor increasing monocytopoiesis. J Exp Med 159:524–536

Sperling U, Kaufmann SHE, Hahn H (1984) Production of macrophage-activating and migration-inhibition factors in vitro by serologically selected and cloned *Listeria monocytogenes*-specific T cells of the Lyt 1 + 2 − phenotype. Infect Immun 46:111–115

Stevenson MM, Kongshavn PAL, Skamene E (1981) Genetic linkage of resistance to *Listeria monocytogenes* with macrophage inflammatory responses. J Immunol 127:402–407

Stevenson MM, Gervais F, Skamene E (1984) Natural resistance to listeriosis: role of host inflammatory responsiveness. Clin Invest Med 7:297–302

Takeya K, Shimotori S, Taniguchi T, Nomoto K (1977) Cellular mechanisms in the protection against infection by *Listeria monocytogenes* in mice. J Gen Microbiol 100:373–379

Tatsukawa K, Mitsuyama M, Takeya K, Nomoto K (1979) Differing contribution of polymorphonuclear cells and macrophages to protection of mice against *Listeria monocytogenes* and *Pseudomonas aeruginosa*. J Gen Microbiol 115:161–166

Unanue ER (1984) Antigen-presenting function of the macrophage. Annu Rev Immunol 2:395–428

Van der Meer C, Hofhuis FMA, Willers JMN (1979) Delayed-type hypersensitivity and acquired cellular resistance in mice immunized with killed *Listeria monocytogenes* and adjuvants. Immunology 37:77–82

Volkman A, Collins FM (1971) The restorative effect of peritoneal macrophages on delayed hypersensitivity following ionizing radiation. Cell Immunol 2:552–566

Von Koenig CH, Finger H, Hof H (1982) Failure of killed *Listeria monocytogenes* vaccine to produce protective immunity. Nature 297:233–234

Zinkernagel RM, Blanden RV, Langman RE (1974) Early appearance of sensitized lymphocytes in mice infected with *Listeria monocytogenes*. J Immunol 112:496–501

Zinkernagel RM, Althage A, Adler B, Blander RV, Davidson WF, Kees U, Dunlop MBC, Shreffler DC (1977) *H-2* restriction of cell-mediated immunity to an intracellular bacterium. Effector T cells are specific for *Listeria* antigen in association with *H-2I* region-coded self-markers. J Exp Med 145:1353–1367

X-Linked Immune Deficiency (*xid*) of CBA/N Mice

L.S. WICKER and I. SCHER

1 Characterization of the *xid* Genetic Defect in CBA/N Mice

1.1 Introduction

The CBA/N mouse strain was bred at the National Institutes of Health from a single litter of a CBA/Harwell mouse. After more than 20 generations of brother-sister mating, the mice derived from this litter, which had normal vigor and reproductive capacity under laboratory conditions, were considered to be inbred. In the early 1970s, it was shown that these mice made abnormal immune responses to polysaccharide antigens (AMSBAUGH et al. 1972; SCHER et al. 1973). Studies of the inheritance of this unresponsiveness demonstrated that F1 male progeny of crosses between CBA/N females and immunologically normal male mice were unresponsive, whereas F1 female progeny were normal. Female and male progeny of crosses between CBA/N males and normal female mice that were heterozygous and hemizygous, respectively, for the X-chromosome of the responsive strain were responders. These data are consistent with an X-linked

Merck Sharp and Dohme Research Laboratories, Rahway, New Jersey 07065, USA

Table 1. Recessive inheritance of *xid* the immune defect[a]

Mouse strain	Source of X-Chromosome	Responses to poly-saccharide antigens
CBA/N		
Male	CBA/N/Y	−
Female	CBA/N/CBA/N	−
DBA/2		
Male	DBA/2/Y	+
Female	DBA/2/DBA/2	+
(CBA/N female × DBA/2 male)F1		
Male	CBA/N/Y	−
Female	CBA/N/DBA/2	+
(DBA/2 female × CBA/N male)F1		
Male	DBA/2/Y	+
Female	DBA/2/CBA/N	+

[a] See text

recessive gene (*xid*) controlling unresponsiveness (Table 1). Recent mapping studies have located the *xid* gene (or group of closely linked genes) between the tabby and hypophosphatemia genes on the X-chromosome (BERNING et al. 1980). The large number of X-linked human immune deficiency diseases (COOPER and SELIGMANN 1977), the association of serum IgM levels with the number X-chromosomes of an individual (ADINOLFI et al. 1978), and the existence of a second mouse strain (DBA/2Ha) with an X-linked immune abnormality (TOMINAGA et al. 1980) suggest that the X-chromosome bears many loci influencing immunity.

1.2 Serum Ig Levels

The level of serum immunoglobulin present in mice can be considered to be in part a measure of their immune function, since circulating antibodies are presumably stimulated by agents which the mice encounter during their development. It is of interest therefore that *xid* mice have low levels of serum IgM (20%) and barely detectable levels of serum IgG3 (15%), when compared with the levels of these isotypes in the serum of control mice (AMSBAUGH et al. 1974; PERLMUTTER et al. 1979). By contrast, the levels of serum IgA, IgG1, IgG2a, and IgG2b are equivalent in *xid* and normal mice.

1.3 Polyclonal T- and B-Cell Activation

T- and B-cell function can be readily assessed by stimulating populations of lymphoid cells with polyclonal activators. While such analysis does not delineate

Table 2. Response of spleen cells from *xid* mice to B- and T-cell mitogens. (SCHER et al. 1975a)

Mitogen	^{3}H-thymidine incorporation (CPM)	
	xid	Normal
Control	190	1153
PHA (0.1%)	17532	21384
ConA (1.25 µg/ml)	76312	94709
LPS (50 µg/ml)	3907	48350
PolyIC (250 µg/ml)	3431	27215

fine immunoregulatory functions or antigen-dependent phenomena, it does provide selective probes for T and B cells. Early studies using the T-cell mitogens phytohemagglutinin (PHA) or concanavalin A (ConA) and spleen cells from *xid* and normal mice showed no significant differences in the resulting proliferative responses (SCHER et al. 1975a). By contrast, the B-cell mitogens lipopolysaccharide (LPS) or polyriboinosinic-polyribocytidilic acid (polyIC) induced proliferative responses in *xid* spleen cell populations which were much reduced when compared with age-matched normal controls (SCHER et al. 1975a) (Table 2). Treatment with anti-Thy-1 plus complement, to enrich for B cells, did not alter the ratio of responsiveness between *xid* and normal spleen cell populations, suggesting that the low responses of the *xid* cells resulted from a primary deficiency in the *xid* B-cells rather than a deficiency in their frequency. It is significant that unlike the macrophages of C3H/HeJ mice (LPS nonresponders), macrophages from *xid* mice produce lymphocyte-activating factor and release prostaglandins in response to LPS (ROSENSTREICH et al. 1978). Thus, mechanisms responsible for the deficiencies in responsiveness of these two mouse strains to LPS must be distinct.

Additional studies of the proliferative responses of *xid* B cells to polyclonal activators indicated that the abnormal responses observed with LPS and polyIC were not unique. Thus, responses of *xid* spleen cells to purified protein derivative or dextran sulfate were considerably reduced when compared with the responses induced by these agents when spleen cells of control mice were studied (FERNANDEZ and MÖLLER, 1977; NARIUCHI and KAKIUCHI 1981). Anti-µ antibodies induced excellent B-cell responses from control adult mice, but failed to induce proliferation when tested with *xid* spleen cells (SIECKMANN et al. 1978; SIECKMANN 1980). The proliferative response of normals was age dependent since spleen cells from 1-week-old normal mice failed to proliferate in response to anti-µ antibodies. Thus, by the criteria of anti-µ-induced proliferation, *xid* adult B-cells resembled the spleen cells of immature normal mice (Table 3).

In vivo or in vitro studies of the LPS-stimulated polyclonal antibody responses of *xid* mice demonstrated a deficiency in the number of plaque-forming cells when compared with control animals (AMSBAUGH et al. 1974; COHEN et al. 1976; SCHER et al. 1977). This was more striking when phenol-extracted

Table 3. Responses of *xid* spleen cells to anti-μ antibodies (SIECKMANN et al. 1978)

Strain	Age weeks	³H-Thymidine Incorporation (CPM)	
		Anti-μ	LPS
(CBA/N × DBA/2)F1 female (normal)	16	126430	115864
(CBA/N × DBA/2)F1 male (*xid*)	16	2097	72846
(C57BL/6 × DBA/2)F1 (normal)	1	2315	11601
	4	13878	45827
	8	108292	74221
	11	121769	79455

preparations of LPS were utilized. Indeed, limiting dilution analysis has indicated that young adult *xid* mice have 50- to 100-fold fewer B cells that will be stimulated to secrete IgM antibody after exposure to phenol-extracted LPS than normal control mice and 10- to 100-fold fewer cells reactive to lipoprotein, an LPS contaminant of preparations prepared by the Boivin procedure (HUBER and MELCHERS 1979).

1.4 Cytotoxic Responses

Spleen cells from *xid* and normal mice sensitized to EL-4 tumor cells in vivo produced equivalent killing of EL-4 target cells in vitro (SCHER et al. 1975a). Normal and *xid* mice also had equivalent capacities to reject allogeneic skin grafts in vivo and their spleen cells were functionally equivalent to those of normals in their capacity to lyse heterologous target cells or chicken red blood cells (SCHER et al. 1975a; NUNN and HERBERMAN 1979). These T- and non-B-cell functions were therefore normal in *xid* mice.

1.5 Responses to Thymic-Independent Antigens

Initial analysis of the immune responses of *xid* mice utilized thymic-independent (TI) antigens. These antigens require little T-cell help in inducing a response, since excellent responses are observed in T-cell-depleted environments such as the nude mouse in vivo or B-cell cultures in vitro. In addition to polyIC and pneumococcal polysaccharide (SIII), the first two polysaccharide antigens studied, *xid* mice are also unresponsive to hapten (i.e., TNP) conjugates of TI antigens such as Ficoll, dextran sulfate, levan, polyvinylpyrrolidone, or polyacrylamide beads (SCHER et al. 1975b; COHEN et al. 1976; FERNANDEZ and MÖLLER 1977; SCHER et al. 1977; LINDSTEN and ANDERSSON 1979; MOND et al. 1979; MORRISSEY et al. 1981). However, the TI antigens TNP-*Brucella abortus* (TNP-BA) or Boivin preparations of TNP-LPS induce excellent anti-TNP responses (MOSIER et al. 1976; MOND et al. 1978; QUINTANS 1979). The differential responsiveness of *xid* mice to these TI antigens forms the basis of a classification

Table 4. Characteristics of TI-1 and TI-2 antigens[a]

	TI-1 (LPS)	TI-2 (Ficoll)
Responses of adult normal mice	+ +	+ +
Responses of adult *xid* mice	+ +	—
Requirement for accessory cells	+ —	+ +
Ability to induce polyclonal B-cell responses	+ +	—

[a] See text

system in which the group of antigens to which *xid* mice do not respond, i.e., TNP-Ficoll, are considered to be TI-2 antigens and the antigens to which they respond, i.e., TNP-LPS, are considered to be TI-1 antigens (MOSIER et al. 1976) (Table 4). That this distinction results from important differences in these TI antigens is supported by the finding that TI-2 antigens have a more stringent requirement for splenic adherent cells (BOSWELL et al. 1980a; MORRISSEY et al. 1981) and T cells (MOND et al. 1980), and that TI-1 and TI-2 antigens stimulate different ratios of antibody isotypes in vivo (SLACK et al. 1980).

1.6 Responses to Thymic-Dependent Antigens

In general, when optimum concentrations of a thymic-dependent (TD) antigen are utilized, the responses of *xid* mice are lower than that of normals (SCHER et al. 1975b; COHEN et al. 1976). In adoptive transfer experiments, abnormal responsiveness to TD antigens occurred when *xid* B cells and either normal or *xid* T cells were used to reconstitute, whereas mice reconstituted with normal B cells and either normal or *xid* T cells gave normal responses (JANEWAY and BARTHOLD 1975; SCHER et al. 1979). Thus, B cells from *xid* mice gave poor or absent IgG anti-SRBC responses in the presence of either *xid* or normal primed T cells. Transfer of increased numbers of *xid* B cells or normal T cells did not improve the responses. Indeed, in vitro experiments studying the primary responses to TNP-KLH have demonstrated that *xid* T cells are comparable to normal T cells in their ability to act as helpers. However, *xid* B cells, in contrast to normal B cells, failed to respond to TNP-KLH in vitro, even in the presence of T helper cells from normal mice (BOSWELL et al. 1980b). These data suggest that the abnormal TD responses of *xid* mice result from a B cell, rather than a helper cell, dysfunction. As is noted later, under certain circumstances *xid* B cells can be driven to make responses to TI-2 antigens. Similarly, under circumstances where T cell help is maximized, as in the splenic focus assay, little difference was noted in the ability of B cells from *xid* and normal mice to make TD responses (METCALF et al. 1980).

1.7 Responses to Phosphocholine

While the concept of TI-1 and TI-2 antigens has proven useful in focusing studies on the mechanisms by which these antigens trigger B cells, it does not

Table 5. TD response (PFCs/spleen) of *xid* and normal mice to PC-KLH (WICKER et al. 1983)

Mouse strain	IgM	IgG3	IgG1	IgG2	IgA
(CBA/N × DBA/2)F1 male (*xid*)	13052	1215	96379	29899	8
%T15	4	8	0	3	9
(CBA/N × DBA/2)F1 female (normal)	30914	5017	92868	13183	11615
%T15	68	49	1	6	85

fully distinguish antigens which initiate responses in *xid* mice. Thus, in early studies *xid* mice apparently failed to respond to phosphocholine (PC) determinants after immunization with either *Ascaris suum* or the C5 variant of *S. pneumoniae* (BROWN et al. 1977; QUINTANS 1977). Indeed, even PC conjugates of TI-1 antigens appeared not to be capable of inducing PC responses in *xid* mice (MOND et al. 1977; QUINTANS and KAPLAN 1978). More recent studies have demonstrated that *xid* mice do make anti-PC responses when challenged with PC coupled to either TI-1 or thymic-dependent antigens (KÖHLER et al. 1981; KENNY et al. 1981; CLOUGH et al. 1981). However, whereas most inbred mice make an anti-PC response which is predominantly of the T15 idiotype, this idiotype is not expressed by *xid* mice during a primary or secondary response (KENNY et al. 1981; CLOUGH et al. 1981). Detailed analysis of T15 idiotype expression in the IgM, IgG3, IgG1, IgG2, and IgA isotypes of normal and *xid* mice responding to PC have revealed an asymmetry in idiotype-isotype expression (WICKER et al. 1982, 1983). Thus, the T15-dominant isotypes in normal mice are IgM, IgG3, and IgA, whereas *xid* mice produce only low levels of these isotypes in response to PC, and these small responses are T15 negative (Table 5).

1.8 Tolerance Induction

The B cells of *xid* mice are more easily made tolerant by hapten-conjugated mouse IgG than B cells of normal mice (METCALF et al. 1979, 1980). Thus, the frequency of IgM anti-DNP clones produced by neonatal or adult *xid* B cells were markedly inhibited if exposed to DNP-IgG before challenge with DNP-KLH. By contrast, DNP-IgG treatment of B cells derived from adult normal mice had little or no influence on the number of anti-DNP clones induced by DNP-KLH.

2 Characteristics of *xid* B Cells

Early studies of *xid* mice indicated that the frequency of their splenic B cells was 25%–35%. In contrast, the frequency of B cells in the spleens of normal mice ranged from 50% to 60% (SCHER et al. 1975a). These findings, along with the inability of *xid* mice to respond to certain TI antigens, suggested that the *xid* gene exerts its influence on immune responses via B cells. This hypothesis was supported by the finding that the helper capacity of T cells from *xid* mice was equivalent to that of normal mice (JANEWAY and BARTHOLD 1975; SCHER et al. 1979; BOSWELL et al. 1980a).

2.1 Surface Immunoglobulin

The unusual character of the B cells of adult *xid* mice has been demonstrated by studies of their surface membrane Ig characteristics (FINKELMAN et al. 1975; SCHER et al. 1976, 1980). The majority of adult splenic B cells of normal mice bear two major Ig isotypes, mu and delta. In the newborn, the surface B-cell IgM density is relatively high, whereas little or no surface IgD is present. As development proceeds, the density of the surface IgM on B cells decreases, coincident with the appearance of increasing amounts of surface IgD, so that in the adult, B-cell surface IgD predominates and surface IgM is present in low density. By contrast, the density of surface IgM remains high on the B cells of adult *xid* mice, so that the majority of *xid* B cells bear high densities of IgM and low to intermediate densities of IgD.

2.2 BLA Determinants

Recently, two B-cell antigens, BLA-1 and BLA-2, have been defined using monoclonal antibodies (HARDY et al. 1983, 1984). BLA-1 appears to be present in similar proportions on *xid* and normal splenic B cells. However, most *xid* splenic B cells bear BLA-2, whereas very few splenic B cells of normal mice express this determinant. Analysis of the frequency of these two markers in developing normal and *xid* mice indicates that BLA-1$^+$, BLA-2$^-$ cells are deficient in *xid* mice, suggesting that this subpopulation may give rise to the predominant B-cell population of normal mice. Moreover, three-color analysis by flow cytometry of the BLA, IgM, and IgD surface markers on B cells demonstrated that *xid* adult B cells were distinct from those of normal immature B cells (despite their similar IgM/IgD ratios), since the latter were BLA-1$^+$ whereas the adult *xid* B cells were BLA-1$^-$. These findings support the view that adult *xid* B cells represent a subpopulation of B cells which is either absent or present in very low numbers in normal mice.

2.3 Minor Lymphocyte-Stimulating and Ia Determinants

Two surface membrane constituents of adult normal B cells appear to be absent, or present in very low quantities, on *xid* B cells, Ia.W39 and the gene products of the minor lymphocyte-stimulating locus (*Mls*). Ia.W39 is defined by an antiserum made by immunizing (CBA/N × C57BL/6)F1 male mice with paternal (C57BL/6) lymphocytes (HUBER et al. 1981). While virtually all IgM-bearing

B cells of normal adults bear this Ia specificity, it is not expressed on *xid*
B cells or on the B cells of neonatal normal mice. The Mls determinants of
murine B cells are studied via the mixed lymphocyte reaction using *H-2* identical
responders and stimulators. Under these circumstances, four codominant *Mls*
alleles have been described, one of which is nonstimulatory. Normal neonatal
B cells do not express a stimulatory capacity in *Mls*-defined MLR until approximately 3–4 weeks of age (AHMED et al. 1977a). Young adult *xid* B cells induce
a very poor or no response (AHMED and SCHER 1976), even when activated
by anti-IgD antibodies (RYAN et al. 1983), a procedure which markedly
augments the *Mls* stimulatory capacity of B cells from normal mice. In
experiments designed to determine if *xid* mice have a "nonstimulatory" *Mls*
allele, it was shown that the *xid* defect prevented the expression of stimulatory
Mls-encoded gene products. Thus spleen cells from (CBA/N × C3H/HeN)F1
female mice induced stimulation indices of approximately 6 from *H-2* identical
CBA/J responder cells, whereas spleen cells from (CBA/N × C3H/HeN) F1 male
xid mice failed to induce stimulation. Studies using old *xid* mice (30 weeks)
demonstrated that the spleen cells of these mice stimulated excellent *Mls*-
determined responses, although these remained considerably lower than the
responses induced by spleen cells from control mice (WEBB et al. 1984). The
expression of Mls determinants by *xid* mice appears, therefore, to be dependent
on the age of the donor mice and not on the state of activation of their B
cells.

2.4 Lyb Antigens

Lyb-3 and Lyb-5 are two different B-cell surface determinants that have been
defined using *xid* mice. Anti-Lyb-3 antiserum [(CBA/N × BALB/c)F1 male anti-
BALB/c] recognizes 50% of adult normal B cells and 15%–20% of neonatal
B cells, but does not react with adult *xid* or newborn B cells (HUBER et al.
1977). Anti-Lyb-3 antiserum has a most interesting capacity to induce anti-
SRBC responses in normal mice after immunization with quantities of SRBC
that otherwise would not be immunogenic. This finding suggested that Lyb-3
might represent a B-cell receptor for T-cell helper factors which are not present
on immature normal or adult *xid* B cells.

Anti-Lyb-5 antiserum is made by extensively absorbing C57BL/6 anti-DBA/
2 serum with lymphoid cells from (CBA/N × DBA/2)F1 *xid* mice (AHMED et al.
1977b). Lyb-5 is defined by the cytotoxic reactivity of anti-Lyb-5 antiserum
with 60% of Ig-bearing splenic B cells of normal mice. By contrast, Lyb-5
determinants are not found on splenic B cells in *xid* mice. Elimination of Lyb-5$^+$
B cells from the spleens of normal mice depletes low-density-bearing surface
IgM B cells (AHMED 1979), TNP-Ficoll-responsive B cells (SUBBARAO et al. 1979),
and all primary PC-specific precursors as measured in the splenic focus assay
(KENNY et al. 1983). Although these data suggest that *xid* B cells represent
the Lyb-5$^-$ B-cell subset from normal mice, it does not prove that the two
cell populations are identical at all biochemical, maturational, and functional
levels. Indeed one recent study by ONO et al. (1983) suggests that normal Lyb-5$^-$

B cells are not identical to *xid* Lyb-5⁻ B cells. Thus, normal B cells treated with anti-Lyb-5 and C′ responded to phenol preparations of LPS, whereas *xid* B cells failed to respond to this stimulus. These data support those of HARDY et al. (1983), who studied the distribution of surface IgM and IgD on BLA⁺ and BLA⁻ B cells of *xid* and normal mice, and concluded that *xid* B cells do not appear to represent a subset of normal B cells.

3 The B Cell as the Primary Cellular Defect in CBA/N Mice

3.1 Reversal of TI-2 Unresponsiveness in *xid* Mice

The absence of TI-2 responses, the abnormal surface phenotype of *xid* B cells, the normal function of *xid* T cells and macrophages, and the generally reduced levels of antibody responsiveness of *xid* mice suggested that the *xid* genetic defect acted primarily, if not totally, at the level of the B cell. The immunodeficiency of *xid* mice does not simply involve the deletion of clones of B cells responsive to determinants on TI-2 antigens since nonresponsiveness to TNP-Ficoll is in contrast to the responsiveness of these mice to the TNP hapten on TI-1 or TD carrier molecules (SCHER et al. 1975a; MOND et al. 1978; METCALF et al. 1980; BOSWELL et al. 1980b). Recent evidence suggests that TI-2-responsive B cells are present in CBA/N mice but that insufficient activation signals are produced by the interaction of antigen with the antigen-binding B cell (Table 6). Indeed, B cells secreting antibody to the TI-2 antigen SIII have been "rescued" from *xid* mice by fusing spleen cells from SIII-immunized *xid* mice with the NS-1 myeloma (SCHROER et al. 1979). Therefore, whereas SIII appears to activate *xid* B cells to a point where they can become successful fusion partners, it does not drive them to terminal differentiation. The failure of *xid* B cells to respond to another TI-2 antigen, TNP-Ficoll, can also be

Table 6. Reversal of TI-2 unresponsiveness in *xid* mice[a]

Mice or cells	Additional stimulus	Response to TI-2 antigens
xid Mice or spleen cells	–	–
xid Mice	8-Bromoguanosine (B-cell adjuvant)	+
xid Mice	Graf-versus-host reaction (T-cell help)	+
xid Mice	T-cell lymphokines (T-cell help)	+
xid Spleen cells	Fusion with myeloma cells	+
xid Bone marrow	Splenic focus system (T-cell help)	+
xid Peyer's patch cells	–	+

[a] See text

reversed with a number of stimuli that probably act on the *xid* B cell following the initial interaction with antigen. It has been found that *xid* mice undergoing a graft-versus-host reaction are responsive to TNP-Ficoll (GOLDING et al. 1982). In vivo responses to TNP-Ficoll can also be restored by injecting *xid* mice with T-cell-derived lymphokines (XUE et al. 1983) or with the B-cell adjuvant 8-bromoguanosine (Wicker, unpublished observation). Finally, in the microenvironment of the Peyer's patches, TNP-Ficoll-responsive B cells have been demonstrated in *xid* mice (ELDRIDGE et al. 1983). The Peyer's patches environment presumably provides unique or additional T-cell-derived signals to *xid* B cells because of the stimulation of T cells by the gut flora (YIKONO et al. 1983). These examples of *xid* B-cell responsiveness to TI-2 antigens indicate that TI-2-responsive B-cell clones are present in *xid* mice but that these clones do not differentiate in the spleens of these mice without additional activation signals. TI-2 B-cell clones of normal mice do not require additional help to respond to these same TI-2 antigens.

In contrast to TNP, PC appears to have few, if any, B-cell precursors in *xid* mice (KENNY et al. 1983). As noted earlier, CBA/N B cells, like neonatal B cells, are highly susceptible to tolerance induction (METCALF et al. 1979, 1980). It is possible that PC-binding B cells of *xid* mice encounter environmental forms of PC and are made tolerant at a time in development when the B cells of normal mice are no longer susceptible to tolerance induction. KLINMAN and STONE (1983) have evidence to support this hypothesis; if *xid* bone marrow cells are placed in an environment such that their first encounter with PC hapten is via a TD antigen in the presence of optimal T-cell help, $T15^+$, PC precursors can be rescued from *xid* mice. As noted for the restoration of TNP-Ficoll responses, the ability to activate $T15^+$, PC precursors in *xid* mice does not imply that the primary *xid* defect has been corrected, more likely, this data suggests that the defect has been bypassed.

The defect in *xid* B cells that results in their unresponsiveness to TI-2 antigens remains unclear. It should be noted that the ability of factors or adjuvants to induce a response to a particular antigen in *xid* mice does not necessarily indicate that these factors are part of the activation pathway in normal mice. These factors may allow for a response to TI-2 antigens in *xid* mice via a different pathway from that used by normal B cells.

3.2 Adoptive Transfer and Chimera Studies

Transfer of anti-Thy $1.2 + C'$-treated normal spleen cells to *xid* mice reconstitutes the ability of such recipients to respond to TNP-Ficoll (SCHER et al. 1975b) and to produce a $T15^+$ anti-PC response (KENNY et al. 1982). These experiments strongly suggest that the primary defect in the *xid* recipient is the B cell and not other cells that are required during the immune response.

To determine if the environment of the *xid* mouse was responsible for the lack of B-cell responsiveness, SCHER et al. (1975b) produced reciprocal bone marrow chimeras between (CBA/N × DBA/2)Fl male and female mice. Female bone marrow matured equally in normal or *xid* irradiated recipients; TNP-Ficoll responses could be elicited from both hosts. Male bone marrow did not provide B cells responsive to TNP-Ficoll, even when it was placed in a normal irradiated recipient. These experiments suggest an intrinsic defect in the stem cell pool in *xid* mice but do not conclusively point to an exclusive B-cell defect since *xid* accessory cells may be defective, which could lead indirectly to the failure of B cells to mature in a normal fashion.

Two lines of experimentation have further supported the hypothesis of an intrinsic B-cell defect in *xid* mice: studies on the cells in normal female mice heterozygous for the *xid* defect and studies of chimeric mice which are the recipients of two sources of bone marrow, *xid* and normal. NAHM et al. (1983) tested female mice heterozygous for two X-chromosome genes, *xid* and *phosphoglycerate kinase-1* (*Pgk-1*). Using *Pgk-1*-encoded isoenzymes as a marker of X-chromosome activtion, it was found that the non-*xid* chromosome was activated in most B cells, whereas both chromosomes were equally represented in non-B-cell lymphocytes and other tissue. Thus, despite the presence of normal T cells and other host cells, B cells expressing the *xid* defect do not mature as do normal B cells. A second line of study also supports the hypothesis of an intrinsic defect in *xid* B cells: *xid* bone marrow cells from CBA/N mice fail to differentiate into PC-responsive B cells even when they mature in a normal host in the presence of normal lymphocytes. (CBA/N × DBA/2)F1-irradiated female mice reconstituted with a mixture of CBA/N and (CBA/N × DBA/2)F1 female bone marrow were immunized with PC-KLH after 3 months. All PC-specific primary PFCs from such chimeras were eliminated by treating spleen cells with an anti-H-2D^d monoclonal antibody and C′ (Wicker and Kenny, unpublished observations). Thus, *xid* B cells do not make a "normal" anti-PC response even though they have matured in a normal recipient and in the presence of normal T, B, and accessory cells. Similar double bone marrow chimeras were produced by SPRENT and BRUCE (1984a). They concluded that *xid* B cells fail to compete in the presence of normal B cells, suggesting that *xid* B cells are abnormal and do not represent a subpopulation of B cells present in normal mice.

3.3 B-Cell Maturation in *xid* Mice

Mice expressing both the nude (gene symbol *nu*) and *xid* genetic defects have been bred to assess the importance of T cells in the maturation of *xid* B cells (MOND et al. 1982, 1983; WORTIS et al. 1982). T-cell dependency has also been studied using adult thymectomized, irradiated bone marrow-reconstituted mice (SPRENT and BRUCE 1984b). In the absence of T cells the *xid* defect is even more pronounced; the lack of T cells and T-cell factors appears greatly to inhibit *xid* B-cell development. This is in sharp contrast with the apparently complete B-cell maturation in normal nude mice. The difference in B-cell function between *nu/nu* and *xid-nu/nu* mice raises the possibility that *xid* B cells are deficient very early in their maturation from the stem cell pool and that these cells are not representative of a subpopulation of normal B cells.

KINCADE (1977) has shown that cells from *xid* mice fail to form B-cell colonies in vitro although the frequency of granulocyte-macrophage progenitors and multipotential stem cells in the bone marrow are normal in these immunodeficient mice. Recently, Wortis et al. (1984) have demonstrated that *xid* B cells also lack the ability to be transformed by Abelson leukemia virus. Lyb-5⁻ B cells derived from CBA/N mice, although functionally similar in many respects to negatively selected Lyb-5⁻ B cells from normal mice, may in fact be quite distinct from these cells. The functional B cells present in *xid* mice may represent a B-cell population which lacks an essential enzyme necessary for normal B-cell maturation. This deficiency presumably occurs early in the B-cell lineage, and the resulting cells, in contrast to normal B cells, are relatively dependent on the presence of factors from T cells to become antigen reactive. These cells never acquire the level of maturation required for responsiveness to TI-2 antigens, or to certain growth or differentiation factors.

The most important issue remains the elucidation of the genetic and resulting biochemical defect present in the *xid* mouse. The definition of the point at which B-cell maturation is hindered in this model will help to define at least one critical step in the B-cell lineage. The *xid* model provides a unique tool for studying the mechanisms of host resistance (Hunter et al. 1979; O'Brien et al. 1980), since *xid* mice fail to respond to many bacterial antigens. In addition, it has been found that autoimmune mice produce significantly reduced levels of autoantibodies when the *xid* genetic defect is bred onto these strains. Important insights into these and other biological issues will be provided by continuing studies of *xid* mice.

References

Adinolfi A, Haddad SA, Seller MJ (1978) X chromosome, complement and serum levels of IgM in man and mouse. J Immunogenet 5:149–156

Ahmed A, Scher I (1976) Studies on non-H-2 linked lymphocyte-activating determinants. II. Nonexpression of Mls determinants in a mouse strain with an X-linked B lymphocyte immune defect. J Immunol 117:1922–1926

Ahmed A, Scher I (1979) Murine B cell heterogeneity defined by anti-Lyb 5, an alloantiserum specific for a late appearing B lymphocyte subpopulation. In: Cooper M, Mosier DE, Scher I, Vitetta ES (eds) B lymphocytes in the immune response. Elsevier, Amsterdam

Ahmed A, Scher I, Sell KW (1977a) Studies on non-H-2 linked lymphocyte-activating determinants. IV. Ontogeny of the Mls product on murine B cells. Cell Immunol 30:122–134

Ahmed A, Scher I, Sharrow SO, Smith AH, Paul WE, Sachs DH, Sell KW (1977b) B lymphocyte heterogeneity: development and characterization of an alloantiserum which distinguishes B lymphocyte differentiation alloantigens. J Exp Med 145:101–110

Amsbaugh DF, Hansen CT, Prescott B, Stashak PW, Barthold DR, Baker PJ (1972) Genetic control of the antibody response to type III pneumococcal polysaccharide in mice. I. Evidence that an X-linked gene plays a decisive role in determining responsiveness. J Exp Med 136:931–936

Amsbaugh DF, Hansen CT, Prescott B, Stashak PW, Asofsky R, Baker PJ (1974) Genetic control of the antibody response to type III pneumococcal polysaccharide in mice II. Relationship between IgM immunoglobulin levels and the ability to give an IgM antibody response. J Exp Med 139:1499–1512

Berning A, Eicher E, Paul WE, Scher I (1980) Mapping of the X-linked immune deficiency mutation (*xid*) of CBA/N mice. J Immunol 124:1875–1877

Boswell HS, Sharrow SO, Singer A (1980a) Role of accessory cells in B cell activation. I. Macrophage presentation of TNP-Ficoll: evidence for macrophage-B cell interaction. J Immunol 124:989–996

Boswell HS, Ahmed A, Scher I, Singer A (1980b) Role of accessory cells in B cell activation.

II. The interaction of B cells with accessory cells results in the exclusive activation of an Lyb5$^+$ B cell population. J Immunol 125:1340–1348

Brown AR, Crandall CA, Crandall RB (1977) The immune response and acquired resistance to *Ascaris suum* infection in mice with an X-linked B lymphocyte defect. J Parasitol 63:950–952

Clough ER, Levy DA, Cebra JJ (1981) (CBA/N × BALB/cJ) F1 male and female mice can be primed to express quantitatively equivalent secondary anti-phosphocholine responses. J Immunol 126:387–389

Cohen PL, Scher I, Mosier DE (1976) In vitro studies of the genetically determined unresponsiveness to thymus-independent antigens in CBA/N mice. J Immunol 116:301–304

Cooper MD, Seligmann M (1977) B and T lymphocytes in immunodeficiency and lymphoproliferative diseases. In: Loor F, Roelants GE (eds) B and T cells in immune recognition, Wiley, New York

Eldridge JH, Kiyono H, Michalek SM, McGhee JR (1983) Evidence for a mature B cell subpopulation in Peyer's patches of young and old *xid* mice. J Exp Med 157:789–794

Fernandez C, Möller G (1977) Immunological unresponsiveness to thymus-independent antigens: two fundamentally different genetic mechanisms of B cell unresponsiveness to dextran. J Exp Med 146:1663–1677

Finkelman FD, Smith AH, Scher I, Paul WE (1975) Abnormal ratio of membrane immunoglobulin classes in mice with an X-linked B lymphocyte defect. J Exp Med 142:1316–1321

Golding H, Foiles PG, Rittenberg MB (1982) Partial reconstitution of trinitrophenylated Ficoll responses and immunoglobulin G3 expression in *xid* mice undergoing graft vs. host reaction. J Immunol 129:2641–2646

Hardy RR, Hayakawa K, Parks DR, Herzenberg LA (1983) Demonstration of B cell maturation in X-linked immunodeficient mice by simultaneous three-color immunofluorescence. Nature 306:270–272

Hardy RR, Hayakawa K, Parks DR, Herzenberg LA, Herzenberg LA (1984) Murine B cell differentiation lineages. J Exp Med 159:1169–1188

Huber B, Melchers F (1979) Frequencies of mitogen-reactive B cells in the mouse. Lipopolysaccharide- lipoprotein- and *Nocardia* mitogen-reactive B cells in CBA/N mice. Eur J Immunol 9:827–829

Huber B, Gershon RK, Cantor H (1977) Identification of a B cell surface structure involved in antigen-dependent triggering: absence of this structure on B cells from CBA/N mutant mice. J Exp Med 145:10–20

Huber BT, Jones PP, Thorley-Lawson DA (1981) Structural analysis of a new B cell differentiation antigen associated with products of the I-A subregion of H-2 complex. Proc Natl Acad Sci USA 78:4525–4529

Hunter KW Jr, Finkelman FD, Strickland GT, Sayles PC, Scher I (1979) Defective resistance to *Plasmodium yoelii* in CBA/N mice. J Immunol 123:133–137

Janeway CA, Barthold DR (1975) An analysis of the defective response of CBA/N mice to T-dependent antigens. J Immunol 115:898–900

Kenny J, Guelde G, Claflin J, Scher I (1981) Altered idiotype response to phosphocholine in mice bearing an X-linked immune defect. J Immunol 127:1629–1633

Kenny JJ, Wicker LS, Guelde G, Scher I (1982) Regulation of T15 idiotype dominance. I. Mice expressing the *xid* immune defect provide normal help to T15$^+$ B cell precursors. J Immunol 129:1534–1538

Kenny JJ, Yaffe LJ, Ahmed A, Metcalf ES (1983) Contribution of Lyb-5$^+$ and Lyb-5$^-$ B cells to the primary and secondary phosphocholine specific-antibody response. J Immunol 130:2574–2579

Kincade PW (1977) Defective colony formation by B lymphocytes from CBA/N and C3H/HeJ mice. J Exp Med 145:249–263

Klinman NR, Stone MR (1983) Role of variable region gene expression and environmental selection in determining the antiphosphorylcholine B cell repertoire. J Exp Med 158:1948–1961

Köhler H, Smyk S, Sung J (1981) Immune responses to phosphorylcholine. VIII. The response of CBA/N mice to PC-LPS. J Immunol 126:1790–1792

Lindsten T, Andersson B (1979) Ontogeny of B cells in CBA/N mice. Evidence for a stage of responsiveness to thymic-independent antigens during development. J Exp Med 150:1285–1292

Metcalf ES, Schrater AF, Klinman NR (1979) Murine models of tolerance induction in developing and mature B cells. Imunol Rev 43:142–183

Metcalf ES, Scher I, Klinman NR (1980) The susceptibility to in vitro tolerance induction of adult B cells from mice with an X-linked B-cell defect. J Exp Med 151:486–491

Mond JJ, Lieberman R, Inman JK, Mosier DE, Paul WE (1977) Inability of mice with a defect in B-lymphocyte maturation to respond to phosphorylcholine on immunogenic carriers. J Exp Med 146:1138–1142

Mond JJ, Scher I, Mosier DE, Blaese M, Paul WE (1978) T-independent responses in B cell-defective CBA/N mice to *Brucella abortus* and to trinitrophenyl (TNP) conjugates of *Brucella abortus*. Eur J Immunol 8:459–463

Mond JJ, Stein KE, Subbarao B, Paul WE (1979) Analysis of B cell activation requirements with TNP-conjugated polyacrylamide beads. J Immunol 123:239–245

Mond JJ, Mongini PKA, Sieckmann D, Paul WE (1980) Role of T lymphocytes in the response to TNP-AECM-Ficoll. J Immunol 125:1066–1070

Mond JJ, Scher I, Cossman J, Kessler S, Mongini PK, Hansen C, Finkelman FD, Paul WE (1982) Role of the thymus in directing the development of a subset of B lymphocytes. J Exp Med 155:924–936

Mond JJ, Norton G, Paul WE, Scher I, Finkelman FD, House S, Schaefer M, Mongini PKA, Hansen C, Bona C (1983) Establishment of an inbred line of mice that express a synergistic immune defect precluding in vitro responses to type 1 and type 2 antigens, B cell mitogens, and a number of T cell-derived helper factors. J Exp Med 158:1401–1414

Morrissey PJ, Boswell HS, Scher I, Singer A (1981) Role of accessory cells in B cell activation. IV. Ia$^+$ accessory cells are required for the in vitro generation of thymic independent type 2 antibody responses to polysaccharide antigens. J Immunol 127:1345–1347

Mosier DE, Scher I, Paul WE (1976) In vitro responses of CBA/N mice: spleen cells of mice with an X-linked defect that precludes immune responses to several thymus-independent antigens can respond to TNP-lipopolysaccharide. J Immunol 117:1363–1369

Nahm MJ, Paslay JW, Davie JM (1983) Unbalanced X chromosome mosaicism in B cells of mice with X linked immunodeficiency. J Exp Med 158:920–931

Nariuchi H, Kakiuchi T (1981) Responses of spleen cells from mice with X-linked B cell defect to polyclonal B-cell activators, purified protein derivative of tuberculin, and dextran sulfate. Cell Immunol 61:375–385

Nunn Me, Herberman RB (1979) Natural cytotoxicity of mouse, rat, and human lymphocytes against heterologous target cells. JNCI 62:765–771

O'Brien AD, Rosenstreich DL, Scher I, Metcalf ES (1980) A model for genetic regulation of innate resistance to bacterial infection: differential sensitivity of inbred mice to *Salmonella typhimurium*. In: Skamene E (ed) Genetic control of natural resistance to infection and malignancy. Academic, New York

Ono S, Yaffe LJ, Ryan JL, Singer A (1983) Functional heterogeneity of the Lyb-5$^-$ B cell subpopulation: mutant *xid* B cells and normal Lyb-5$^-$ B cells differ in their responsiveness to phenol-extracted lipopolysaccharide. J Immunol 130:2014–2021

Perlmutter RM, Nahm M, Stein KE, Slack J, Zitron I, Paul WE, Davie JM (1979) Immunoglobulin subclass-specific immunodeficiency in mice with an X-linked B-lymphocyte defect. J Exp Med 149:993–998

Quintans J (1977) The "patchy" immunodeficiency of CBA/N mice. Eur J Immunol 7:749–751

Quintans, J (1979) The immune response of CBA/N mice and their F1 hybrids to 2,4,6-trinitrophenyl-ated (TNP) antigens. I. Analysis of the response to TNP-coupled lipopolysaccharide in vivo and at the clonal level. Eur J Immunol 9:67–71

Quintans J, Kaplan RB (1978) Failure of CBA/N mice to respond to thymus-dependent and thymus-independent phosphorylcholine antigens. Cell Immunol 38:294–298

Rosenstreich DL, Vogel SN, Jacques A, Wahl LM, Scher I, Mergenhagen SE (1978) Differential endotoxin sensitivity of lymphocytes and macrophages from mice with an X-lined defect in B cell maturation. J Immunol 121:685–690

Ryan JL, Mond JJ, Finkelman FD, Scher I (1983) Enhancement of the mixed lymphocyte reaction by in vivo treatment of stimulator spleen cells with anti-IgD antibody. J Immunol 130:2534–2541

Scher I, Frantz MD, Steinberg AD (1973) The genetics of the immune response to a synthetic double-stranded RNA in a mutant CBA/N mouse strain. J Immunol 110:1396–1401

Scher I, Ahmed A, Strong DM, Steinberg AD, Paul WE (1975a) X-linked B lymphocyte immune defect in CBA/HN mice. I. Studies of the function and composition of spleen cells. J Exp Med 141:788–803

Scher I, Steinberg AD, Berning AK, Paul WE (1975b) X-linked B-lymphocyte immune defect in CBA/N mice. II. Studies of the mechanisms underlying the immune defect. J Exp Med 142:637–650

Scher I, Sharrow SO, Wistar R, Asofsky R, Paul WE (1976) B-lymphocyte heterogeneity: ontogenetic development and organ distribution of B-lymphocyte populations defined by their density of surface immunoglobulin. J Exp Med 144:494–506

Scher I, Zaldivar NM, Mosier DE (1977) B-lymphocyte subpopulations and endotoxin responses in CBA/N mice. In: Schlesinger D (ed) Microbiology – 1977. American Society for Microbiology, Washington DC

Scher I, Berning AK, Asofsky R (1979) X-linked B lymphocyte defect in CBA/N mice. IV. Cellular and environmental influences on the thymus dependent IgG anti-sheep red blood cell response. J Immunol 123:477–486

Scher I, Berning AK, Kessler S, Finkelman FD (1980) Development of B lymphocytes in the mouse; studies of the frequency and distribution of surface IgM and IgD in normal and immune-defective CBA/N F1 mice. J Immunol 125:1686–1693

Schroer KR, Kim KJ, Prescott B, Baker PJ (1979) Generation of anti-type III pneumococcal polysaccharide hybridomas from mice with an X-linked B-lymphocyte defect. J Exp Med 150:698–702

Sieckmann DG (1980) The use of anti-immunoglobulins to induce a signal for cell division in B lymphocytes via their membrane IgM and IgD. Immunol Rev 52:55–74

Sieckmann DG, Scher I, Asofsky R, Mosier DE, Paul WE (1978) Activation of mouse lymphocytes by anti-immunoblobulin II. A thymus-independent response by a mature subset of B lymphocytes. J Exp Med 148:1628–1643

Slack J, Der-Balian GP, Nahm M, Davie JM (1980) Subclass restriction of murine antibodies. II. The IgG plaque-forming cell response to thymus-independent type 1 and type 2 antigens in normal mice and mice expressing an X-linked immunodeficiency. J Exp Med 151:853–862

Sprent J, Bruce J (1984a) Physiology of B cells in mice with X-linked immunodeficiency. II. Influence of the thymus and mature T cells on B cell differentiation. J Exp Med 160:335–340

Sprent J, Bruce J (1984b) Physiology of B cells in mice with X-linked immunodeficiency (*xid*). III. Disappearance of *xid* B cells in double bone marrow chimeras. J Exp Med 160:711–723

Subbarao B, Mosier DE, Ahmed A, Mond JJ, Scher I, Paul WE (1979) Role of a nonimmunoglobulin cell surface determinant in the activation of B lymphocytes by thymus-independent antigens. J Exp Med 149:495–506

Tominaga A, Takatsu K, Hamaoka T (1980) Antigen-induced T cell-replacing factor (TRF). II. X-linked gene control for the expression of TRF-acceptor site(s) on B lymphocytes and preparation of specific antiserum to that acceptor. J Immunol 124:2423–2429

Webb SR, Mosier DE, Wilson DE, Sprent T (1984) Negative selection in vivo reveals expression of strong Mls determinants in mice with X-linked immunodeficiency. J Exp Med 160:108–115

Wicker LS, Guelde G, Scher I, Kenny JJ (1982) Antibodies from the Lyb-5⁻ B cell subset predominate in the secondary IgG response to phosphocholine. J Immunol 129:950–953

Wicker LS, Guelde S, Scher I, Kenny JJ (1983) The asymmetry in idiotype-isotype expression in the response to phosphocholine is due to divergence in the expressed repertoires of Lyb-5⁺ and Lyb-5⁻ B cells. J Immunol 131:2468–2476

Wortis HH, Burkly L, Hughes D, Roschelle S, Waneck G (1982) Lack of mature B cells in nude mice with X-linked immune deficiency. J Exp Med 155:903–913

Wortis HH, Karagogeos D, Rosenberg N (1984) Cell differentiation in nude/X-linked immune-deficient mice. J Cell Biochem [Suppl] 8A:139

Xue B, Bell MK, Thorbecke GJ (1983) Influence of lymphokines on the anti-TNP-Ficoll response of normal and X-linked B lymphocyte-defective mice. J Immunol 131:1698–1701

Yikono H, Mosteller LM, Eldridge JH, Michalek SM, McGhee JR (1983) Immunoglobulin A responses in *xid* mice. Oral antigen primes Peyer's patch cells for in vitro immune responses and secretory antibody production. J Immunol 131:2616–2622

Genetic Control of the Susceptibility
to Pneumococcal Infection*

D.E. Briles[1], J. Horowitz[1], L.S. McDaniel[1], W.H. Benjamin, Jr.[1],
J.L. Claflin[2], C.L. Booker[1], G. Scott[1], and C. Forman[1]

1 Introduction

As with other pathogens used to infect mice, it has been known for some
time that mouse strains differ in their ability to be infected by *Streptococcus
pneumoniae* (WEBSTER 1933; SCHULTZ et al. 1936; RAKE 1936). These observations provide evidence that there are genetic differences in mice that affect
their resistance to pneumococci. Some of the earliest studies of genetic differences in pathogenesis of *S. pneumoniae* infections were done with the BS and
BR mice which were bred by Lesslie Webster to be susceptible or resistant to

---

* This work has been supported by NIH grants CA 16673, CA 13148, AI 15986, AI 18557, and
AI 21548 and Alabama Research Institute grant 420. David E. Briles is the recipient of a Research
Career Development Award, AI 00498.

[1] Cellular Immunobiology Unit of the Tumor Institute, Department of Microbiology, and the Comprehensive Cancer Center, University of Alabama at Birmingham, Birmingham, Alabama 35294,
USA
[2] Department of Microbiology and Immunology, University of Michigan Medical School, Ann
Arbor, Michigan 48109, USA

Table 1. Susceptibility of mice to pneumococcal infection

Mice	LD_{50}[a]	
	i.n.	i.v.
BSVS	10^3	10^2
CBA/J	$>10^7$	10^3
BALB/cByJ	10^7	10^3

[a] Inoculation with type 3 strain EF 10197 by i.n. (intranasal) or i.v. (intravenous) routes

infection with *Salmonella enteritidis* (WEBSTER 1933). When Webster infected these mice with pneumococci by the nasal route, he found that BS mice were more susceptible to pneumococcal infection than the BR mice (WEBSTER 1933).

BS mice do not exist today, but their descendents, the BSVS mice which were bred to be susceptible to both *S. enteritidis* and St. Louis encephalitis virus (WEBSTER 1933), are still highly susceptible to intranasal pneumococcal infection with certain pneumococcal strains (Table 1). The relationship, if any, between the susceptibility of BSVS mice to salmonella and pneumococcal infection is not clear. Webster bred the BSVS mice from an originally outbred population using resistance to oral salmonella infection as his selection parameter. It is plausible that in the development of BSVS mice, certain allele(s) affecting the entry of bacteria from mucosal sites may have been selected for, which could affect the acquisition of both of these agents from oral and intranasal routes.

Whatever gene is responsible for the increased susceptibility of BSVS mice to intranasal infections with *S. pneumoniae* (WEBSTER 1933; RAKE 1936; SCHULTZ et al. 1936), it is apparently different from the Ity^s allele of the Ity locus which is thought to be responsible for much of the susceptibility of BSVS mice to salmonella (O'BRIEN et al. 1980; BRILES et al. 1981a). Whether the route of salmonella infection is oral or i.v., the Ity^s allele has been shown to cause higher rates of accumulation of salmonella in the liver and spleen of infected Ity^s versus Ity^r mice (WEBSTER 1933; HORMAECHE 1979; O'BRIEN et al. 1980). It seems unlikely that the Ity locus affects the susceptibility of mice to pneumococcal infection, since when mice are infected with *S. pneumoniae* i.v., rather than orally, the Ity^r mouse strains, BRVR and A/J, are no more resistant to pneumococcal infection than the Ity^s strains, BSVS and BALB/cJ (BRILES et al. 1981a).

In other studies, we have examined the susceptibility of a number of different inbred mouse strains to fatal i.v. infection with particular isolates of type 3 and type 1 pneumococci (BRILES et al. 1983). The data obtained reveal a number of additional differences in the pneumococcal susceptibility of inbred mice (Fig. 1). The two pneumococcal isolates tested showed varying degrees of virulence in different inbred mouse strains, indicating that the mouse strains differ in genes for resistance to pneumococcal infection. Since some strains of mice,

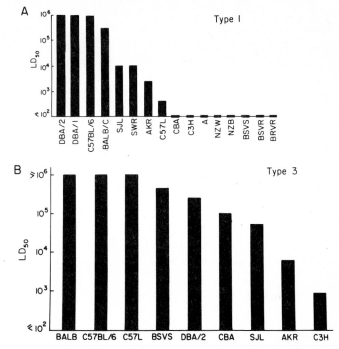

Fig. 1 A, B. LD$_{50}$ of *S. pneumoniae*, strain 1.1 for various inbred mouse strains. Data for infections with type 1 strain 1.1 are shown in **A**. Data for infections with type 3 strain WU2 are shown in **B**

such as BSVS, showed very different susceptibilities to the two pneumococcal strains, it appears that the pneumococcal strains differ in their virulence genes and that interactions between the pneumococcal and murine genotypes may be important in the pathogenesis of the pneumococcal infections.

In an effort to learn more about the genes controlling the susceptibility to pneumococcal infection, the type 1 and type 3 strains were used to infect the B × H and AK × L recombinant inbred (RI) mice respectively. The B × H strains were derived from C57BL/6J and C3H/HeJ parents and the AK × L strains from AKR/J and C57L/J parents. The median days alive, calculated as the reciprocal mean, for each RI strain are shown in Fig. 2. Within each of the two experiments, the variation in the resistance of the individual strains is consistent with the inheritance of resistance by genes at multiple loci.

Using the statistical approach described by Stocker, this volume, we have compared the resistance and susceptibility of these strains with the known distributions of parental alleles within each set of RI strains. In the case of the B × H mice, we failed to detect any clear associations. In the case of the AK × L mice infected with WU2 *S. pneumoniae*, we have observed that the chromosome 1 *Akp-1* allele of C57L mice is strongly associated with resistance. The rank order value for this comparison is 2.62, indicating that the association between *Akp-1* and resistance would have occurred by chance less than once in 1/100 observations. Because of the large number of comparisons made, how-

A

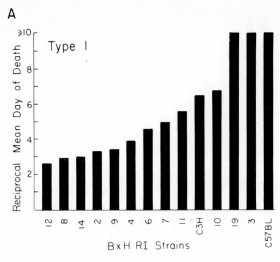

B

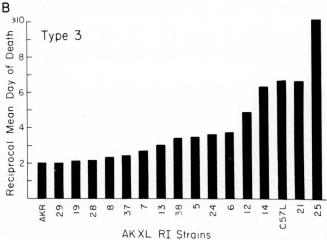

Fig. 2 A, B. Reciprocal mean days alive of recombinant inbred mice infected with *S. pneumoniae*. A shows data from BXH RI strains were infected with 10^4 type 1, strain 1.1

ever, we cannot be certain that the association of *Akp-1* with resistance is not due to chance without an independent genetic confirmation (see Stocker, this volume).

2 Pneumococcal Infections in Mice with *Lps^{d/d}*, *nu/nu*, and *xid* Defects

We have also examined the effects of known immunodeficiency defects on the susceptibility to pneumococcal infection. Neither the *nu/nu* nor the *Lps^{d/d}* defects led to increased susceptibility to i.v. infection with type 1 pneumococci (Table 2).

Table 2. Lack of effect of *nu* and *Lps^d* on susceptibility to type 1 *S. pneumoniae*[a]

Mice	Genotype	Number of mice	Median days alive	% Alive at 20 days
SWISS	*nu/+*	6	9.7	33
SWISS	*nu/nu*	6	8.8	50
C3H/HeN	*Lps^n/Lps^n*	23	3.1	14
C3H/HeJ	*Lps^d/Lps^d*	28	5.1	15

[a] Infected i.v. with 10^2 *S. pneumoniae*

Table 3. Comparison of virulence and ability to bind anti-PC antibody

Streptococcus pneumoniae		LD$_{50}$ in mice				% ^{125}I-anti-PC bound[b]	% Remaining in blood after 4 h with anti-PC[c]
Strain	Type	CBA/N	CBC[a] (males)	CBC (females)	BALB/cJ		
1.1	1	10^1	2×10^5	$> 10^6$	$> 10^6$	14	38
D-39	2	10^1	10^1	$> 10^6$	$> 10^6$	44	10
WU2	3	10^1	10^1	$> 10^6$	10^5	6	44
A66	3	10^1	10^1	10^3	10^2	10	33
EF10197	3	10^1	10^2	10^3	10^4	36	17
3296	4	10^1	–	–	10^1	8	28
5.2	5	10^1	10^1	10^1	10^1	7	61

[a] (CBA/N × BALB/cJ)F$_1$, (CBC) males express the *xid* defect of CBA/N mice. BALB/c and CBC female mice are immunologically normal

[b] Binding of ^{125}I-IgM anti-PC antibody 22.1A4 to 10^7 live pneumococci

[c] 10^6 *S. pneumoniae* injected i.v. into CBC male mice that received 20 μg IgG$_3$ anti-PC antibody 1 h earlier. The percentage is calculated versus CBC male mice receiving an equal dose of *S. pneumoniae* but no antibody

However, mice expressing the *xid* phenotype were found to be highly susceptible to pneumococcal infection with several serotypes of *S. pneumoniae*. From the data in Table 3 it can be seen that the dose of pneumococci at which 50% of the animals survive (LD$_{50}$) is generally lower in *xid* than non-*xid* mice (see also BRILES et al. 1980; BRILES et al. 1981c; YOTHER et al. 1982b). One exception to this trend is with the type 5 isolate, which was so highly virulent in non-*xid* mice that the detection of additional virulence in *xid* mice was precluded (YOTHER et al. 1982b).

The effect of other genes on the susceptibility caused by the *xid* genotype was apparent when we examined the susceptibility of *xid* mice to our strain 1.1 type 1 isolate; a rather interesting observation was made. When CBA/N (xid) and CBA/J mice were compared, the high susceptibility of the CBA background (Fig. 1) permitted little or no additional susceptibility due to *xid* to be visualized. When this experiment was repeated with (CBA/N × BALB/c)F$_1$ *xid* males and females (Table 3) or (CBA/N × DBA/2)F$_1$ *xid* males and females, we found that the BALB/c and DBA/2 resistance genes were dominant in the cross and largely abrogated any susceptibility caused by the *xid* locus. In the CBA/N ×

Table 4. Effect of DBA/2 and *xid* genes on resistance to strain 1.1 type 1 *S. pneumoniae*[a]

Mice	n	% Alive at			Median days alive
		2 days	10 days	17 days	
CBA	9	0	0	0	2
CBA/J	17	65	6	6	4
(CBA/N × DBA/2)F$_1$ males	10	80	70	30	13
(CBA/N × DBA/2)F$_1$ females	9	100	100	89	>17
DBA/2	9	100	100	100	>17

[a] Infected i.v. with 10^4 strain 1.1 *S. pneumoniae*. Median survival calculated as reciprocal mean days alive

DBA/2 cross, the greater susceptibility of *xid* than normal F$_1$ mice was not seen, however, until 10 days post inoculation when the *xid* animals began to die of pneumococcal infection (Table 4).

Thus for this particular type 1 isolate, the effect of normal and defective alleles of the *xid* locus appear to be largely overshadowed by other as yet undefined resistance genes of DBA/2 and BALB/cJ mice. Thus in spite of the implication of the data in Fig. 1, there must be a major difference in the pathogenesis of the type 1 and WU2 strains.

2.1 Mechanism of Action of the *xid* Defect

The major defect in *xid* mice is thought to be associated with the differentiation of their B cells (Kongshavn, this volume). This defect is most pronounced with a subset of antigens referred to as type 2 antigens. Most of these antigens are polysaccharides and are frequently thymus independent (TI-2) (SCHER 1982), although some of them (BRILES et al. 1982c; PRESS 1981; ROSENWASSER and HUBER 1981), such as streptococcal group A carbohydrate, have been shown to require the presence of T cells for optimal responsiveness (TD-2). Thus it seemed likely that the high susceptibility of *xid* mice to pneumococcal infection (BRILES et al. 1981c) was because they failed to make normal antibody responses to carbohydrate antigens (SLACK et al. 1980; SCHER 1982; BRILES et al. 1982c).

The *xid* trait has been shown to increase the susceptibility of mice to a number of other bacterial, viral, and parasitic agents (BROWN et al. 1977; HUNTER et al. 1979; O'BRIEN et al. 1979; LEVY et al. 1984; MARQUIS et al. 1985). In each case, the increased virulence of these pathogens in *xid* mice is most likely an indication that antibodies, probably anticarbohydrate antibodies, play a protective role in the disease processes of these agents.

2.1.1 Natural Antibody

Since *xid* mice infected with as few as ten colony-forming units of many of the pneumococcal strains survived no longer than 36 h, it was unlikely that the non-*xid* mice were protected by their ability to make anticarbohydrate immune responses to the infecting bacteria (BRILES et al. 1981c; YOTHER et al. 1982b). The most plausible way the *xid* trait could have affected protection on such a short time scale was if the normal, but not the *xid*, mice had preexisting antipneumococcal antibody in their serum prior to the infection (BRILES et al. 1981c). This appeared to be the case, since passive transfer of normal serum from non-*xid* mice could protect *xid* mice from fatal pneumococcal infection (BRILES et al. 1981c).

Pneumococcal capsules have been shown to be a major virulence factor of the pneumococcus (GRIFFIN 1928; MACLEOD and KRAUS 1950; WOOD and SMITH 1949), and antibodies against the capsular polysaccharides have been shown to be highly effective at mediating protection against pneumococcal infection (WHITE 1938; MACLEOD et al. 1945; AUSTRIAN 1979). Thus it seemed likely that the pneumococcal antigens recognized by the protective natural serum antibodies would be capsular polysaccharides.

2.2 Protective Anti-PC Antibody in Mice

However, the protective antibodies in the normal serum of mice did not turn out to be reactive with pneumococcal capsular polysaccharide since absorption of the normal serum with either encapsulated or unencapsulated pneumococci removed the protective capacity of the normal serum (BRILES et al. 1981c). The protective antibodies were found to be specific for the phosphocholine (PC) residues of pneumococcal teichoic acids and F antigen (TOMASZ 1967; BRUNDISH and BADDILEY 1968; BRILES and TOMASZ 1973), since absorption of the normal serum with phenyl-PC-Sepharose but not with phenyl-*N*-acetyl-glycosaminide-Sepharose removed the protective capacity of the serum (BRILES et al. 1981c). Antibodies to the PC determinant of these pneumococcal cell wall polysaccharides have been intensively studied (COHN et al. 1969; CLAFLIN and DAVIE 1975; ANDRES et al. 1981; PERLMUTTER et al. 1984). These antibodies are readily elicited by immunization with killed unencapsulated pneumococci (COHN et al. 1969). These antibodies appear to constitute a major portion of the mouse antipneumococcal response, since inhibition with free PC greatly reduces the antipneumococcal antibody titer of immune sera to rough pneumococci by over 90-fold.

The identity of the antigens that elicit these naturally occurring anti-PC antibodies is unknown. It seems likely that they are antigens from the normal flora since germ-free mice do not make detectable levels of anti-PC antibody but begin making anti-PC antibodies after conventionalization (LIEBERMAN et al. 1974).

The importance of anti-PC antibody in the protection of mice from pneumococcal infection was also apparent from studies where antibody to the T15 idiotype of anti-PC antibodies was used to suppress the anti-PC response of adult BALB/c mice (BRILES et al. 1981c). When this was done, the mice were much more susceptible to pneumococcal infection than mice of the same inbred strain that were not idiotype suppressed. In another study (MCNAMARA et al. 1984), have shown that by immunizing mice with anti-T15 idiotype they can increase the production of anti-PC antibodies, which lends to an increase in the resistance of the mice to pneumococcal infection (McNamara et al. 1984). In other studies, it has been shown that when anti-PC antibodies were elicited in mice either by immunization with killed rough pneumococci or with PC-KLH, the mice show increased resistance to fatal infection with *S. pneumoniae* (Briles et al. 1982a; Wallick et al. 1983; McNamara et al. 1984), although, when rough pneumococci were used for immunization, it is not clear that the only antipneumococcal antibodies contributing to protection are those reactive with PC (see Sect. 6.1).

The ability of anti-PC antibodies to protect mice from pneumococcal infection has also been demonstrated directly by passively transferring hybridoma anti-PC antibodies to *xid* and normal mice (Briles et al. 1981c; Yother et al. 1982b; Szu et al. 1983).

Anti-PC antibodies and PC-binding human C-reactive protein (see Sect. 6.2) have been shown to prevent fatal mouse infection caused by human pneumococcal isolates of serotypes 1, 2, 3, 4, and 6A (Mold et al. 1981; Briles et al. 1981c; Yother et al. 1982a; Yother et al. 1982b; Szu et al. 1983). This finding is consistent with the fact that all pneumococci have PC in both their lipoteichoic acids (F-antigen) and in their cell wall teichoic acids (a portion of the cell wall C-carbohydrate) (Brundish and Baddiley 1968; Briles and Tomasz 1973).

The fact that anti-PC antibodies could provide protection against pneumococcal infection was a surprise since it had been thought that the capsules of pneumococci would effectively block any effector functions of the immune system that might be mediated by anti-cell wall antibodies. When we measured the binding of radioactive IgM anti-PC antibody to live encapsulated and unencapsulated pneumococci, we found that it took 300 times as many encapsulated as unencapsulated pneumococci to bind the same amount of antibody (Yother et al. 1982b). Thus the capsule did a very defective job of hiding most of the cell wall. However, the small amount of the cell wall components that were still exposed permitted the binding of enough anti-PC antibodies to facilitate the in vivo destruction of the pneumococci (Briles et al. 1981c).

Even though anti-PC antibodies could be shown to bind to all pneumococcal isolates tested (Table 3 and Yother et al. 1982b), it failed to protect mice against all of them. For example, anti-PC antibody could not protect against infection with particular type 3, 4, 5 and 6A isolates (Yother et al. 1982b; Szu et al. 1983; Briles, unpublished). The reason for the resistance of these isolates to protection by anti-PC antibodies is unknown. It appears, however, that the relative ability of anti-PC antibody to bind the different pneumococcal isolates is not associated with the virulence of the isolate or their ability to resist clearance by anti-PC antibody (Table 3). Furthermore, it appears that high virulence is not necessarily associated with resistance to blood clearance by anti-PC antibodies (Table 3).

2.3 "Naturally Occurring" Anti-PC Antibody in Man

Other species have also been shown to have detectable levels of natural anti-PC antibody in their serum (Claflin and Davie 1974). In humans, we have observed

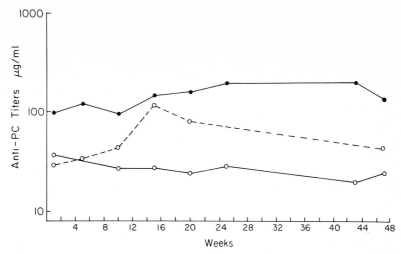

Fig. 3. Fluctuation of anti-PC levels over time for three individuals

that 95% of adults less than 60 years of age have levels of anti-PC antibody greater than 15 µg/ml (BRILES and SCOTT 1985). In children less than 6 months of age (GRAY et al. 1983) and in about one-third of adults over 60 years of age (BRILES and SCOTT 1985), there are undetectable levels of serum anti-PC antibody. In the children, there is a gradual conversion of anti-PC responsiveness until about 2 years of age, at which time most of the children have titers that are even higher than normal adult levels (GRAY et al. 1983). The poor anti-PC responsiveness in infants is similar to what has been observed for other polysaccharide antigens, where it has been found that polysaccharides are poorly immunogenic until the children reach about 2 years of age (PELTOLA et al. 1977; COWAN et al. 1978; GRAY et al. 1983).

From a prospective study with 30 children, it appears that the very high titers in some of the children may result in part from immune responses they have made to their frequent pneumococcal otitis media infections (GRAY et al. 1983). This conclusion is consistent with recent studies of older children where we have shown that children recovering from bacterial pneumonia have higher anti-PC levels than children recovering from viral pneumonia (GRAY, SCOTT, DILLON and BRILES 1985, unpublished observation).

When the anti-PC levels of ten adults were monitored for a period of 1–3 years, it was found that the level of anti-PC antibodies in the serum of each individual remained relatively constant and that each person had their own basal anti-PC level that varied in individual subjects from 10 to 200 µg/ml (SCOTT and BRILES 1985). Results from three of these individuals are shown in Fig. 3. Even though titers significantly higher or lower than the basal level were occasionally observed in some individuals, their serum anti-PC levels always returned to their individual basal level within a few weeks (SCOTT and BRILES 1985).

The mechanisms regulating anti-PC antibody levels in adults are unknown, but the constancy of their serum levels suggests that the basal level of anti-PC antibody is probably not dependent on the irregular instances of pneumococcal infection or carriage. It seems more likely that, as in the case of the mouse, environmental antigens, possibly from the normal flora, may account for the natural immunization.

At present, there are no data indicating that human anti-PC antibody is protective against pneumococcal infection in humans. However, there are some interesting associations between anti-PC levels and incidence of pneumococcal disease. It has been observed that those children between 6 and 48 months of age with the lowest serum anti-PC levels are either those with the least carriages of pneumococci or those with the most pneumococcal disease (GRAY et al. 1983). One interpretation of these data is that the children with low anti-PC levels but little or no carriage of pneumococci have not had enough pneumococcal exposure to be well immunized against pneumococcal PC. Those children with multiple pneumococcal carriages but low anti-PC levels may have had a higher carriage rate because their antibody responses to anti-PC, or to carbohydrates in general, have not yet matured. This association between high carriage and low anti-PC levels could be only because anti-PC levels might be serving as an indicator of responsiveness to capsular carbohydrates, and not because they play a role in protection against pneumococci.

2.4 Structure of Anti-PC Antibodies

The observation that anti-PC antibodies are protective against pneumococcal infection has provided an excellent system to study antibody-mediated protection against gram-positive infections. Mouse anti-PC antibodies are perhaps the most studied groups of antibodies. The genes specifying these antibodies have been described and sequenced (PERLMUTTER et al. 1984). A large number of hybridoma antibodies to PC have been produced (ANDRES et al. 1981; GEARHART et al. 1981). Many of these hybridoma antibodies have been characterized in terms of their fine specificity, idiotypy, and sequence. Virtually all anti-PC antibodies utilize the same V_H, D_H, and J_H genes.

Mouse anti-PC antibodies can be classified into three familes, T15, M603, or M511, depending on whether they use light chain genes from the V_k8, V_k22, or V_k24 families (PERLMUTTER et al. 1984). It has been shown that each of the three families of anti-PC antibodies have characteristic specificity and idiotypy (ANDRES et al. 1981; CLAFLIN et al. 1981). Within the T15 family in particular, there is very close uniformity of the specificity and amino acid sequence of the different anti-PC antibodies (ANDRES et al. 1981; CLAFLIN et al. 1981; GEARHART et al. 1981). Because of the extreme V-region uniformity of the anti-PC antibodies, they have been ideal for comparing the differences in protective capacity of antibacterial antibodies of different isotypes (BRILES et al. 1981b; BRILES et al. 1984a).

3 Relative Protective Effects of Antibodies of Different Isotypes

3.1 Comparison of IgM, IgG$_3$, and IgA

A comparison of antibodies of T15 IgM, IgG$_3$, and IgA anti-PC antibodies has revealed that IgM and IgG$_3$, but not IgA, antibodies are protective against fatal infection following i.v. administration of pneumococci (BRILES et al. 1981b). The IgG$_3$ antibodies were observed to be 10–100 times as effective as IgM anti-PC antibodies (BRILES et al. 1981b). This result is independent of whether or not the antibody is administered i.p. prior to injection of the bacteria or administered i.v. along with the inoculum. We have also shown that the blood levels of passive IgM and IgG antibody at 1-h postinjection are independent of whether or not the antibody is injected i.p. or i.v. (McDANIEL et al. 1984b).

When *xid* mice were protected with IgG or IgM anti-PC antibodies, we observed that 20 µg pf IgG antibodies were able to protect against 10^5 times higher levels of pneumococci than even 100 µg IgM anti-PC antibodies (Table 5). Poor protective efficacy of the IgM as compared with IgG antibody was a little surprising since IgM antibody would be expected to be much better at mediating complement fixation, a process that should enhance phagocytosis and killing by phagocytes. Since it has more than two binding sites, IgM antibody should also be more avid than an IgG antibody with the same variable regions. This expectation was borne out when IgG and IgM anti-PC antibodies were used to inhibit the binding of a radioactive IgM anti-PC antibody to live rough pneumococci in solution. We found that IgM anti-PC antibody was at least 30 times as effective at inhibiting the binding of the labeled anti-PC antibody as IgG.

The reasons for the more effective protection with IgG than with IgM antibodies are not yet clear. It may be due to the fact that IgG but not IgM is recognized by Fc receptors on phagocytes. It may also be related in part to the fact that about one-fifth as much of the injected IgM as compared with IgG antibody reaches the circulation and its half-life in our studies was about

Table 5. Relative protective capacities of IgM and IgG anti-PC antibodies[a]

Anti-PC antibody	LD$_{50}$ WU2 *S. pneumoniae*
None	$<10^2$
20 µg IgM	10^2
100 µg IgM	10^3
20 µg IgG^3	3×10^7
20 µg IgG^1	$>10^8$
20 µg IgG$_{2b}$	$>10^8$

[a] Antibody given i.p. 1 h prior to i.v. injection of *S. pneumoniae*

11 h, as compared with 32 h for IgG_3 (MᶜDᴀɴɪᴇʟ et al. 1984b). In an effort to compensate for the shorter half-life of IgM, we have injected it every 8 h for a total of five injections. When this is done, the LD_{50} of the pneumococci still did not exceed 10^4, barely exceeding the protection caused by a single injection of IgM antibody.

3.2 Comparison of the Protective Effects of IgG_1, IgG_{2b}, and IgG_3 Anti-PC Antibodies

Most, but not all, of the IgG anticarbohydrate antibodies produced by the mouse have been reported to be of the IgG_3 isotype (Pᴇʀʟᴍᴜᴛᴛᴇʀ et al. 1978; Dᴇʀ Bᴀʟɪᴀɴ et al. 1980; Sᴛᴇɪɴ et al. 1983). It seems likely that these IgG_3 antibodies are made by the subpopulation of B cells that are absent in *xid* mice (Sᴄʜᴇʀ 1982).

We had speculated that the reason for the isotype restriction of anticarbohydrate antibodies might be so that antibodies with the isotype most effective at mediating protection against bacteria would be produced (Bʀɪʟᴇꜱ et al. 1981b). However, when we determined the protective effects of T15 IgG_1, IgG_{2b}, and IgG_3 anti-PC antibodies, we found that antibodies of the three isotypes were indistinguishable in their protective capacity (Bʀɪʟᴇꜱ et al. 1984a).

This raised the possibility that the restriction of IgG anticarbohydrate antibodies to the IgG_3 isotype might be because these antibodies were better at facilitating complement-mediated lysis of gram-negative bacteria. Since there was not an appropriate gram-negative bacterium with PC on its surface, we measured the ability of anti-PC antibodies to promote complement-mediated lysis of PC-conjugated rabbit red blood cells.

When this was done, we observed that although IgG_3 anti-PC antibody could promote the lysis of PC-conjugated rabbit red blood cells in the presence of fresh normal mouse serum, it was no more effective at this, on a weight basis, than IgG_1 or IgG_{2b} anti-PC antibody (Fig. 4) (Bᴏᴏᴋᴇʀ, Cʟᴀꜰʟɪɴ and Bʀɪʟᴇꜱ 1985, unpublished observation).

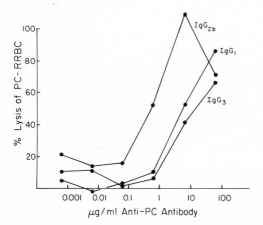

Fig. 4. Concentration of IgG anti-PC antibody required to mediate lysis of PC-rabbit red blood cells in the presence of fresh *xid* mouse serum

While the possibility remains that IgG_3 may be specialized for other functions such as antibody-dependent cell-mediated cytotoxicity (ADCC), we have speculated that the reason virtually all species have multiple IgG subclasses and that virtually all antigens exibit antigenic restriction of their response to a subset of the IgG subclasses is to allow isotype-specific (HOOVER and LYNCH 1983) regulation to play a role in modulating antibody responses (BRILES et al. 1984a).

4 Comparison of the Comparative Protective Effects of Anti-PC Antibodies of Different Idiotypes

The clear differences in the specificity of anti-PC antibodies of the three idiotypic families (ANDRES et al. 1981; CLAFLIN et al. 1981; YOUNG et al. 1985) prompted us to determine whether or not there would be a correlation between the protective effects of the antibodies and the idiotype family they belonged to. We examined this by comparing the protective capacity of IgM and IgG_1 anti-PC antibodies from the different idiotype families. Within each isotype, we found that the T15 anti-PC antibodies were the only ones that provided significant protection (BRILES et al. 1982b; BRILES et al. 1984b). This finding was interesting from two points of view. The fact that some of the T15 antibodies, such as the IgM M2, appear to be germline in sequence (PERLMUTTER et al. 1984) indicates that germline antibodies are not only protective against pneumococcal infection but also are more protective than any other anti-PC antibodies we have examined (BRILES et al. 1982b; BRILES et al. 1984b). The other impressive thing about this result is that there is enough uniformity within the anti-PC antibody families to allow the individual antibodies to be characterized as protective or nonprotective based on their idiotypes. If this turns out to be the case in human immune responses to bacterial antigens, it should be possible to determine whether a vaccine is eliciting protective antibodies in a human population before it is subjected to extensive trials.

The observation that germline antibodies can protect against bacterial infections has prompted us to speculate that many of the germline immunoglobulin V region genes may exist in their present form because they have been selected to protect against certain common infections or groups of infections (BRILES et al. 1984b). This could be particularly important for the relatively thymus independent antigens, such as bacterial polysaccharides, which apparently fail to elicit responses that undergo affinity maturation as occurs for thymus-dependent responses (KIMBALL 1972; BRILES and DAVIE 1975).

The failure of anticarbohydrate antibody responses to undergo affinity maturation could prevent them from being able to take advantage of somatic mutation in the generation of high-affinity antibodies. Thus, the only way an animal could be assured of making a protective anticarbohydrate antibody to any particular antigen would be to have germline sequence coding for the protective antibodies. If this were the case, a large portion of the germline V region genes might be able to participate in antibody responses to pathogens common in the evolutionary history of a species. Whether this prediction is true or false will only become apparent after a great deal more is learned about the specificity and structure of anticarbohydrate antibodies.

5 Mechanism of the Anti-PC Antibodies

It is assumed that the protective effects of anti-PC antibodies are mediated by their ability to promote phagocytosis, which leads to the intracellular killing of the pneumococci by macrophages and polymorphonuclear leukocytes (PMNs). The evidence for this is indirect. The requirement for phagocytes is suggested by the fact that at 37° C in the presence of anti-PC antibody and fresh mouse serum, pneumococci are not killed. In fact, they can be grown under these conditions, and their net growth rate is no less in fresh serum than in heat-inactivated serum. The ability of pneumococci to grow in the presence of anti-PC antibody also rules out the possibility that anti-PC antibody has a direct deleterious effect on the pneumococci, such as possibly interfering with cell wall synthesis or blocking the acquisition of an essential nutrient.

Complement fixation appears to be necessary for the full protective effects of both IgM and IgG$_3$ anti-PC antibody since mice whose C3 has been depleted with cobra venom factor are not protected against infection by IgM or IgG$_3$ anti-PC antibody (BRILES and FORMAN 1985).

It has been possible to demonstrate that IgM and IgG anti-PC antibodies lead to the blood clearance of pneumococci with the maximum effect at 4 h post i.v. injection of bacteria. The doses of antibody required for optimal blood clearance are only slightly higher than the minimal doses of antibody required to protect mice from fatal pneumococcal infection (McDANIEL et al. 1984b). The cleared bacteria are apparently killed since they do not accumulate as living organisms in the reticuloendothelial organs such as the liver, spleen, lung, or kidney (McDANIEL et al. 1984b).

6 *xid* Mouse Provides a Convenient Test System with Which to Study Other Mediators of Antipneumococcal Protection

One of the important uses of immunodeficient animals is that they provide a test environment in which to measure the effects of various protective functions whose effects might be difficult to sort out in the presence of other modes of protection. Since *xid* mice lack naturally occurring serum anti-PC antibody, they provide a system in which antipneumococcal antibodies or other mediators of opsonization can be tested in the absence of the normally protective levels of naturally occurring anti-PC antibodies (BRILES et al. 1981c).

6.1 Protective Effects of Antibodies to Pneumococcal Surface Proteins

We have produced a panel of hybridoma antibodies reactive with pneumococcal surface proteins that appear to recognize at least two different molecular species of about 70 000 and 40 000 mw. Several of the antibodies to the 70 000-mw proteins are able to protect mice from fatal infection with certain strains of pneumococci (McDANIEL et al. 1984a). Although these proteins are susceptible to hydrolysis by pepsin and trypsin, they resist denaturation by boiling at pH 2, a property that they share with streptococcal M proteins (AUSTRIAN and MAC-LEOD 1949; SWIFT et al. 1943).

Electrophoretic and immunochemical analysis of the pneumococcal proteins detected by these antibodies indicate that these proteins are highly polymorphic. Although we have not detected an antigenic determinant that is common to all of the proteins of either group, cross-reactions among the proteins are common (McDANIEL et al. 1984a, McDANIEL et al. 1985).

6.2 Protective Effects of C-Reactive Protein

Human C-reactive protein (CRP) is produced in humans subsequent to pneumococcal infection, other bacterial infection, inflammation, and tissue damage. It has been reported to opsonize some species of bacteria including pneumococci (KINDMARK 1977) and to activate the classical complement pathway when binding to the appropriate ligand (KAPLAN and VOLANAKIS 1974). Human CRP has been shown to be able to protect normal mice from $S.$ $pneumoniae$ infection by increasing the LD_{50} by five fold (MOLD et al. 1981; YOTHER et al. 1982a). Similarly, we have recently shown that in xid mice infected with $S.$ $pneumoniae$, treatment with CRP increased the LD_{50} by ten fold (from one to ten bacteria). Both human and rabbit CRP are highly effective at mediating blood clearance of pneumococci. Even when xid mice were challenged with 10^5 times the LD_{50} of $S.$ $pneumoniae$, CRP and rabbit C-reactive protein (C × RP) significantly prolonged the life of the animals, although it rarely results in complete protection despite markedly enhanced bacterial clearance. It seems that this effect of CRP is mediated, at least partly, via activation of the complement system since treatment with cobra venom factor largely abolishes the prolonged survival and bacterial blood clearance mediated by CRP and C × RP. These results make it clear that the acute phase proteins, CRP and C × RP, may play an important protective role as a major part of the early defense mechanism against bacterial infections. Human and possibly other CRPs appear to provide protection during the critical time prior to the beginning of specific antibody production.

References

Andres CM, Maddalena A, Hudak S, Young NM, Claflin JL (1981) Anti-phosphocholine hybridoma antibodies. II. Functional analysis of binding sites within these antibody families. J Exp Med 154:1584–1598

Austrian R (1979) Pneumococcal vaccine: development and prospects. Am J Med 67:547–549

Austrian R, MacLeod CM (1949) A type specific protein from pneumococcus. J Exp Med 89:439–450

Briles DE, Davie JM (1975) Clonal dominance. I. Restricted nature of the IgM antibody response to Group A streptococcal carbohydrate in mice. J Exp Med 141:1291–1307

Briles DE, Forman C (1985) Complement is required for the blood clearance of *S. pneumoniae* by IgG as well as IgM anti-PC antibodies. (Manuscript in preparation.)

Briles DE, Scott G (1985) Naturally occurring anti-phosphorylcholine levels in normal humans. (Manuscript in preparation.)

Briles EB, Tomasz A (1973) Pneumococcal Forssman antigen: a choline-containing lipoteichoic acid. J Biol Chem 248:6394–6397

Briles DE, Nahm M, Schroer K, Baker P, Davie J (1980) Susceptibility of (CBA/N × DBA/2)F$_1$ male mice to infection with type 3 (*Streptococcus pneumoniae*. In: Skamene E (ed) Perspectives in immunology: genetic control of natural resistance to infection and malignancy. Academic, London, pp 173–177

Briles DE, Benjamin WH, Curtis AW, Davie JM (1981a) A genetic locus responsible for salmonella susceptibility in BSVS mice is not responsible for the limited T-dependent immune responsiveness of BSVS mice. J Immunol 127:906–911

Briles DE, Claflin JL, Schroer K, Forman C (1981b) Mouse IgG$_3$ antibodies are highly protective against infection with *Streptococcus pneumoniae*. Nature 294:88–90

Briles DE, Nahm M, Schroer K, Davie J, Baker P, Kearney J, Barletta R (1981c) Anti-phosphocholine antibodies found in normal mouse serum are protective against intravenous infection with type 3 *S. pneumoniae*. J Exp Med 153:694–705

Briles DE, Barletta R, Nahm M, Schroer K, Baker P, Kearney J (1982a) Use of hybridoma technology to study anti-pneumococcal antibodies: anti-phosphocholine antibodies can protect mice against infection with type 3 *Streptococcus pneumoniae*. In: Robbins JB, Hill JC, Sadoff JC (ed) Seminars in infectious diseases, vol IV: bacterial vaccines. Thieme-Stratton, New York, pp 1–5

Briles DE, Forman C, Hudak S, Claflin JL (1982b) Anti-PC antibodies of the T15 idiotype are optimally protective against *Streptococcus pneumoniae*. J Exp Med 156:1177–1185

Briles DE, Nahm M, Marion TN, Perlmutter RM, Davie JM (1982c) Streptococcal group-A carbohydrate has properties of both a thymus-independent (TI-2) and a thymus-dependent antigen. J Immunol 128:2032–2035

Briles DE, Forman C, Benjamin WH, Yother J (1983) Pathogenesis of type 1 *Streptococcus pneumoniae* in mice. Fed Proc 42(4):861

Briles DE, Forman C, Hudak S, Claflin JL (1984a) The effects of subclass on the ability of IgG anti-phosphocholine antibodies. J Mol Cell Immunol 1:305–309

Briles DE, Forman C, Hudak S, Claflin JF (1984b) The effects of idiotype on the ability of IgG$_1$ anti-phosphocholine antibodies to protect mice from fatal infection with *Streptococcus pneumoniae*. Eur J Immunol 14:1029–1030

Brown AR, Crandall CA, Crandall RB (1977) The immune response and acquired resistance to *Ascaris suum* infection in mice with an X-linked B lymphocyte defect. J Parasitol 63:950–952

Brundish DE, Baddiley J (1968) Pneumococcal C-substance, a ribitol teichoic acid containing choline phosphate. Biochem J 110:573–582

Claflin JL, Davie JM (1974) Clonal nature of the immune response to phosphorylcholine. III. Species-specific characteristics of rodent anti-phosphorylcholine antibodies. J Immunol 113:1678–1684

Claflin JL, Davie JM (1975) Clonal nature of the immune response to phosphorylcholine (PC) V. J Exp Med 141:1073–1083

Claflin JL, Hudak S, Maddalena A (1981) Anti-phosphocholine hybridoma antibodies. I. Direct evidence for three distinct families of antibodies in the murine response. J Exp Med 153:352–364

Cohn M, Notani G, Rice S (1969) Characterization of the antibody to the C-carbohydrate produced by a transplantable mouse plasmacytoma. Immunochemistry 6:111–123

Cowan MJ, Ammann AJ, Wara DW, Howie VM, Schultz L, Doyle N, Kaplan M (1978) Pneumococcal polysaccharide immunization in infants and children. Pediatrics 62:721–727

Der Balian GP, Slack J, Clevinger BL, Bazin H, Davie JM (1980) Subclass restriction of murine antibodies. III. Antigens that stimulate IgG$_3$ in the mouse stimulate IgG$_{2c}$ in the rat. J Exp Med 152:209–218

Gearhart PJ, Johnson ND, Douglas R, Hood L (1981) IgG antibodies to phosphorylcholine exhibit more diversity than their IgM counterparts. Nature (London) 291:29–34

Gray BM, Dillon HC, Briles DE (1983) Epidemiological studies of *Streptococcus pneumonia* in infants: development of antibody to phosphocholine. J CLin Microbiol 18:1102–1107

Griffin FJ (1928) The significance of pneumococcal types. J Hyg 27:113–159

Hoover RG, Lynch RG (1983) Isotype-specific suppression of IgA: suppression of IgA responses in BALB/c mice by Tα cells. J Immunol 130:521–523

Hormaeche CE (1979) Natural resistance to *Salmonella typhimurium* in different inbred mouse strains. Immunology 137:311–318

Hunter KW Jr, Finkelman FD, Strickland GT, Sayles PC, Scher I (1979) Defective resistance to *Plasmodium yoelii* in CBA/N mice. J Immunol 123:133–137

Kaplan MH, Volanakis JE (1974) Interaction of C-reactive protein complexes with the complement system. I. Consumption of human complement associated with the reaction of C-reactive protein with pneumococcal C-polysaccharide and with the choline phosphatides lecithin and sphingomyelin. J Immunol 112:2135–2147

Kimball JW (1972) Maturation of immune response to type III pneumococcal polysaccharide. Immunochemistry 9:1169–1184

Kindmark CO (1977) Stimulating effect of C-reactive protein on phagocytosis of various species of pathogenic bacteria. Clin Exp Immunol 8:941–948

Levy L, Aizer F, Bejar C, Lutsky I, Mor N (1984) Experimental myco-bacterial infections of CBA/N mice. Isr J Med Sci 20:598–602

Lieberman R, Potter M, Mushinski EB, Humphrey W Jr, Rudikoff S (1974) Genesis of a new IgV$_H$ (T15 idiotype) marker in the mouse regulating natural antibody to phosphorylcholine. J Exp Med 139:983–1001

MacLeod CM, Krauss MR (1950) Relation of virulence of pneumococcal strains for mice to the quality of capsular polysaccharide formed in vitro. J Exp Med 92:1–9

MacLeod CM, Hodges RG, Heidelberger M, Bernhard WG (1945) Prevention of pneumococcal pneumonia by immunization with specific capsular polysaccharides. J Exp Med 82:445–465

Marquis G, Montplaisir S, Pelletier M, Mousseau S, Auger P (1985) Genetic resistance to murine cryptococcosis: increased susceptibility in the CBA/N *xid* mutant strain of mice. Infect Immun 47:282–287

McDaniel LS, Scott G, Kearney JF, Briles DE (1984a) Monoclonal antibodies against protease sensitive pneumococcal antigens can protect mice from fatal infection with *Streptococcal pneumoniae*. J Exp Med 160:386–397

McDaniel LS, Benjamin WH Jr, Forman C, Briles DE (1984b) Blood clearance by anti-phosphocholine antibodies as a mechanism of protection in experimental pneumococcal bacteremia. J Immunol 133:3308–3312

McDaniel LS, Scott G, Widenhofer K, Briles DE (1986) Analysis of a surface protein of *Streptococcus pneumoniae* recognized by protective monoclonal antibodies. Microbiol Pathogenesis. In press.

McNamara MK, Ward RE, Kohler H (1984) Monoclonal idiotype vaccine against *Streptococcus pneumoniae* infection. Science 226:1325–1326

Mold CS, Nakayama S, Holzer TJ, Gerwurz H, DuClos TW (1981) C-reactive protein is protective against *Streptococcus pneumoniae* infection in mice. J Exp Med 154:1703–1708

O'Brien AD, Scher I, Campbell GH, MacDermott RP, Formal SB (1979) Susceptibility of CBA/N mice to infection with *Salmonella typhimurium*. J Immunol 123:720–724

O'Brien AD, Rosenstreich DL, Taylor BA (1980) Control of natural resistance to *Salmonella typhimurium* and *Leishmania donovani* in mice by closely linked but distinct genetic loci. Nature 287:440–442

O'Brien AD, Rosenstreich DL, Metcalf ES, Scher I (1980) Differential sensitivity of inbred mice to *Salmonella typhimurium*: A model for genetic regulation of innate resistance to bacterial infection. In: Skamene E (ed) Perspectives in immunology: Genetic control of natural resistance to infection and malignancy. Academic, London, pp 101–114

Peltola H, Kayhty H, Sivonen A, Makela PH (1977) *Haemophilus influenzae* type b capsular polysaccharide vaccine in children. Pediatrics 60:730–737

Perlmutter RM, Hansburg D, Briles DE, Nicolotti RA, Davie JM (1978) Subclass restriction of murine anti-carbohydrate antibodies. J Immunol 121:566–572

Perlmutter RM, Crews ST, Douglas R, Sorensen G, Johnson N, Nivera N, Gearhart PJ, Hood L (1984) The generation of diversity in phosphorylcholine-binding antibodies. Adv Immunol 35:1–59

Press JL (1981) The CBA/N defect defines two classes of T cell-dependent antigens. J Immunol 126:1234–1240

Rake G (1936) Pathology of pneumococcus infection in mice following intranasal instillation. J Exp Med 63:17–31

Rosenwasser LJ, Huber BT (1981) The *xid* gene controls Ia.W39-associated immune response gene function. J Exp Med 153:1113–1123

Scher I (1982) The CBA/N mouse strain: an experimental model illustrating influences of the X-chromosome on immunity. Adv Immunol 33:1–71

Schultz H, Gorer PA, Finlayson MH (1936) The resistance of four mouse lines to bacterial infection. J Hyg 36:37–49

Slack J, Der Balian GP, Nahm M, Davie JM (1980) Subclass restriction of murine antibodies. II. The IgG plaque-forming cell response to thymus-independent type 1 and type 2 antigens in normal mice and mice expressing an X-linked immunodeficiency. J Exp Med 151:853–862

Stein KE, Zopf DA, Miller CB, Johnson BM, Mongini PKA, Ahmed A, Paul WE (1983) Immune response to a thymus-dependent form of B512 Dextran requires the presence of Lyb-5$^+$ lymphocytes. J Exp Med 157:657–666

Swift HF, Wilson AT, Lancefield RC (1943) Typing group A hemolytic streptococci by M precipitin reactions in capillary pipettes. J Exp Med 78:127–133

Szu SC, Clarke S, Robbins JB (1983) Protection against pneumococcal infection in mice conferred by phosphocholine-binding antibodies: specificity of the phosphocholine binding and relation to several types. Infect Immun 39:993–999

Tomasz A (1967) Choline in the cell wall of a bacterium: novel type of polymer-linked choline in pneumococcus. Science 157:694–697

Wallick S, Claflin JL, Briles DE (1983) Resistance to *Streptococcus pneumoniae* is induced by a phosphocholine-protein conjugate. J Immunol 130:2871–2875

Webster LT (1933) Inherited and acquired factors in resistance to infection. J Exp Med 57:819–843

White B (1938) The biology of *Pneumococcus*. Oxford University Press, London, pp 58–612

Wood B, Smith MR (1949) The inhibition of surface phagocytosis by the capsular slime layer of pneumococcus type III. J Exp Med 90:85–95

Yother J, Volanakis JE, Briles DE (1982a) Human C-reactive protein is protective against fatal *Streptococcus pneumoniae* infection in mice. J Immunol 128:2374–2376

Yother J, Forman C, Gray B, Briles DE (1982b) Protection of mice from infection with *Streptococcus pneumoniae* by anti-phosphocholine antibody. Infect Immun 36:184–188

Young MN, Williams RE, Claflin JL (1985) The circular dichroism of phosphocholine-specific mouse hybridoma and myeloma proteins: unusual properties of the hybridoma protein 101.6G6. Mol Immunol 22:305–311

Lps Gene Regulation of Mucosal Immunity and Susceptibility to *Salmonella* Infection in Mice

D.E. Colwell, S.M. Michalek, and J.R. McGhee

1 Historical Discovery of LPS-Hyporesponsive Mice and the *Lps* Gene

In 1940, Hill and coworkers reported that the inoculation of a heterogeneous population of mice with endotoxin, followed by the selective interbreeding of survivors, produced a strain of mice which displayed significantly increased resistance to the lethal effects of endotoxin. This finding suggested that the degree of sensitivity to endotoxin challenge was a heritable trait in mice. The demonstration by Sultzer (1968) that inbred C3H/HeJ mice were resistant to the endotoxin-induced infiltration of mononuclear cells into the peritoneal cavity, as well as to the toxicity of endotoxin, identified this mouse strain as a valuable tool for the study of mechanisms of resistance to endotoxin toxicity and genetic control of murine sensitivity to endotoxin. By analyzing the LPS responsiveness of 11 different mouse strains which were derived from the C3H strain between 1920 and 1968, Glode and Rosenstreich (1976) traced the chronology of mutational events which led to the defect in LPS responsiveness of C3H/HeJ mice. The unresponsiveness of C3H/HeJ mice to LPS was reported to be caused by a genetic mutation which occurred some time between 1960 and 1965 and was inbred into the C3H/HeJ line by 1968 (Glode and Rosenstreich 1976; reviewed by Vogel et al. 1979).

Department of Microbiology, University of Alabama at Birmingham, Birmingham, Alabama 35294, USA

The numerous in vivo and in vitro studies which followed have revealed that C3H/HeJ mice are not only highly resistant to the toxicity of endotoxin, but also are refractory to other biological effects of purified endotoxin or its protein-free lipopolysaccharide (LPS) component. These LPS-induced effects to which C3H/HeJ mice are hyporesponsive include B-cell mitogenicity, polyclonal B-cell activation, adjuvanticity, macrophage activation, stimulation of nonspecific resistance to bacterial infection, and induction of serum amyloid A-stimulating factor, glucocorticoid-antagonizing factor, and interferon production (reviewed by MORRISON and RYAN 1979; and SCIBIENSKI 1981).

In contrast, C3H/HeJ mice are immunologically normal in many other respects, such as in their ability to mount an immune response to the polysaccharide portion of LPS, in the ability of their serum complement to be activated by either the classical or alternative pathway following interaction with LPS, and in the ability of their B cells to be activated by non-LPS mitogens such as dextran sulfate, polyriboinosinic acid-polyribocytidylic acid, and endotoxin protein (reviewed by MORRISON and RYAN 1979; and SCIBIENSKI 1981). In addition, MUSSON et al. (1978) reported that while intravenously injected LPS was found associated with spleen cells of LPS-responsive C3H/St mice to a greater extent than with spleen cells of LPS-hyporesponsive C3H/HeJ mice, there was no significant difference between the rates of clearance of LPS from the circulation of these two mouse strains. Taken together, these results suggest that the defect in the C3H/HeJ mouse strain is manifested at the cellular level.

1.1 Identification and Location of the *Lps* Gene

Our understanding of the genetic control of murine sensitivity to bacterial LPS was advanced by studies which analyzed the patterns of LPS responsiveness of F1 hybrid progeny from crosses between LPS-hyporesponsive C3H/HeJ and a number of different LPS-responsive mouse strains, and the patterns of responsiveness of backcross progeny from matings between F1 hybrid mice and the LPS-responsive or LPS-hyporesponsive parent. One early study reported that the mitogenic response of spleen cells from (C3H.SW-Ig-1 [b] × C3H/HeJ)F1 mice to LPS was comparable to that of the C3H.SW-Ig-1 [b] parent, suggesting dominant inheritance of LPS responsiveness (WATSON and RIBLET 1974); however, subsequent studies using other LPS-responsive parental strains have reported that F1 hybrid mice exhibited intermediate levels of responsiveness to several biological and immunological effects of LPS, including toxicity, immunogenicity, adjuvanticity, and mitogenicity (McGHEE et al. 1979; reviewed by WATSON et al. 1978a; and MORRISON and RYAN 1979). Similarly, other studies have reported an intermediate degree of intraperitoneal inflammation and level of serum interferon in F1 hybrid mice following the injection of LPS (McGHEE et al. 1979; reviewed by MORRISON and RYAN 1979). Autoradiographic studies by KELLY and WATSON (1977) indicated that half as many cells in a spleen cell culture from F1 hybrid mice as in a culture from the LPS-responsive parental strain were mitogenically stimulated by LPS, suggesting that the intermediate level of LPS responsiveness of F1 mice resulted either from allelic exclusion

Linkage Group VIII
On Chromosome 4

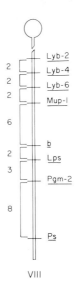

Lps^n/Lps^n Full LPS Responses

Fig. 1. Diagrammatic representation of the location of the *Lps* gene on chromosome 4 of the mouse

Lps^n/Lps^d Intermediate LPS Responses

Lps^d/Lps^d Low LPS Responses

or a gene dosage phenomenon. Examination of the LPS responsiveness of (C3H/HeJ × LPS-responder) F1 mice backcrossed with the C3H/HeJ low-responder or LPS-high-responder parent revealed a 1:1 ratio of low-responder to intermediate-responder, or intermediate-responder to high-responder progeny, respectively, with no observed linkage to the sex, allotype, or H-2 haplotype of the animals studied (reviewed by MORRISON and RYAN 1979; and WATSON et al. 1980). In addition, the backcross analyses demonstrated concordant segregation of the expression of immunogenic, adjuvant, polyclonal, and mitogenic responses to LPS (WATSON et al. 1980). Although these results were originally interpreted to mean that responsiveness to LPS is under polygenic control (SULTZER 1972; GLODE and ROSENSTREICH 1976), it is now generally accepted that a single autosomal gene is responsible for the defective responsiveness of C3H/HeJ mice to LPS, and that alleles of this gene are codominantly expressed (reviewed by SCIBIENSKI 1981).

The chromosomal location of the gene responsible for unresponsiveness to LPS in the C3H/HeJ mouse strain was identified by WATSON and colleagues (1977). Using recombinant inbred strains of mice produced by inbreeding pairs of offspring from the F2 generation of a cross between LPS-responsive C57BL/6J and LPS-hyporesponsive C3H/HeJ parents, these investigators demonstrated that the inheritance of responsiveness to LPS in the recombinant inbred mice segregated concordantly with the phenotypic expression of the C57BL/6J *Mup-1*[b] allele of the *Mup-1* locus, which controls electrophoretic migration patterns of the major urinary protein. Unresponsiveness to LPS segregated with expression of the *Mup-1*[a] allele of the C3H/HeJ mouse strain. Linkage of LPS responsiveness to the expression of the *Mup-1* locus suggested that the gene encoding responsiveness to LPS in the mouse was located on chromosome 4, since *Mup-1* was linked to the brown coat color locus (*b*) on chromosome 4 (WATSON et al. 1977). Backcross linkage analysis of the inheritance of LPS responsiveness and *Mup-1* in (C3H/HeJ × C57BL/6J)F1 × C3H/HeJ mice confirmed the coordinate inheritance of LPS responsiveness and *Mup-1* expression and, thus, verified the location of the LPS response gene, the alleles of which have been termed *Lps*[d] (for defective responsiveness) and *Lps*[n] (for normal responsiveness), in the VIII linkage group on chromosome 4 of the mouse (WATSON et al. 1978b, 1978c). Further genetic recombination studies have mapped the *Lps* locus to a position very close to the *b* locus, between the *Mup-1* and *Ps* (polysyndactyly) loci (WATSON et al. 1978a) (Fig. 1).

1.2 Mouse Strains Other than the C3H/HeJ Which are Hyporesponsive to LPS

In addition to the hyporesponsiveness of C3H/HeJ mice to LPS, defective LPS-responses have been described in three other strains of mice, the CBA/N, whose X-linked defect appears to be limited to the B-cell population (reviewed by MORRISON and RYAN 1979; and SCHER 1981), the C57BL/10ScCR (COUTINHO et al. 1977), and the C57BL/10ScN (nu/nu) (VOGEL et al. 1979) strains. The finding of LPS responsiveness which resulted from genetic complementation in F1 female offspring from crosses between CBA/N and C3H/HeJ parents demonstrated that the hyporesponsiveness of B cells from these strains of mice to mitogenic stimulation by LPS was determined by different genetic loci (GLODE and ROSENSTREICH 1976). When the relationship between the inheritance of LPS hyporesponsiveness in C57BL/10ScCR and C3H/HeJ mice was examined, no responsiveness to LPS, and thus no genetic complementation, was observed in F1 hybrid progeny derived from crosses between these two strains (COUTINHO and MEO 1978). This suggested that C57BL/10ScCR mice carry a defective allele at the same *Lps* locus responsible for LPS hyporesponsiveness in C3H/HeJ mice. The demonstration that (C57BL/10ScCR × LPS-responder) F1 mice exhibited high levels of responsiveness to LPS and that (F1 hybrid × C57BL/10ScCR) backcross mice segregated into high and low responders (COUTINHO and MEO 1978) indicated that hyporesponsiveness to LPS in C57BL/10ScCR mice was inherited as an autosomal recessive trait. In contrast, the alleles encoding LPS responsiveness in (C3H/HeJ × LPS-responder) F1 mice were found to be codominantly expressed, resulting in an intermediate level of responsiveness (COUTINHO et al. 1975; reviewed by SCIBIENSKI 1981). These findings suggested that the LPS gene defects in the C57BL/10ScCR and C3H/HeJ mouse strains were unique. In summary, one can conclude that the C57BL/10ScCR strain, its progenitor strain C57BL/10ScN (VOGEL et al. 1979), and the C3H/HeJ strain all possess an *Lps*[d]/*Lps*[d] genotype which is responsible for LPS hyporesponsiveness.

2 *Lps* Gene Regulation of IgA Immune Responses and Oral Tolerance

Lipopolysaccharide induces a wide range of measurable effects on lymphoreticular cells, e.g., macrophages, and B- and T-lymphocytes. One may, for convenience, describe these LPS effects as either immunostimulatory or suppressive (Table 1; reviewed by MORRISON and RYAN 1979; JIRILLO et al. 1984b). Biostimulatory effects of LPS include B-cell mitogenicity and polyclonal B-cell activation; however, only one-third of murine B cells are triggered to divide and differentiate into immunoglobulin (Ig)-secreting plasma cells in response to LPS (ANDERSON et al. 1977). LPS is immunogenic and small amounts will elicit antibody responses to the various LPS regions. One of the first well-documented immunostimulatory effects of LPS was its adjuvant property (JOHNSON et al. 1956), i.e., administration of appropriate doses of LPS with T-cell-dependent (TD) antigen resulted in enhanced immune responses. LPS activates macrophages and induces the production of a variety of mediators including interleukin 1 (IL 1) (an endogenous pyrogen) and the others listed in Table 1. There is evidence that such mediators elicit toxic effects including fever and even death of the host. LPS may also suppress lymphoid cell responses and administration of LPS prior to a TD antigen can lead to complete suppression of immune responses to that antigen. Endogenous LPS has marked effects on lymphoreticular cells in the gut and strongly influences induction of IgA responses and oral tolerance. This will be discussed in further detail below. All

Table 1. LPS effects on lymphoreticular cells

I. Immunostimulation

  A. B-cell mitogenicity
  B. Polyclonal B-cell activation
  C. Immunogenicity
  D. Adjuvancy
  E. Triggering of small T-cell subpopulations
  F. Enhancement of lectin-induced T-cell responses
  G. Macrophage activation and mediator production
    1. Interleukin 1 (IL 1) or endogenous pyrogen (EP)
    2. Colony-stimulating factors (CSFs)
    3. Interferon (IFN)
    4. Glucocorticoid-antagonizing factor (GAF)
    5. Tumor necrosis factor (TNF)
    6. Numerous others

II. Immunosuppression

  A. Endotoxin tolerance (host insensitivity to LPS)
  B. Suppression of responses to T-dependent antigens
  C. Regulation of oral tolerance and IgA responses

III. Host toxicity

  A. Pyrogenicity
  B. Abortion?
  C. Generalized Schwartzman reaction
  D. Lethality

LPS effects listed in Table 1 are defective in Lps^d/Lps^d mice, i.e., the C3H/HeJ, C57BL/10ScN, and C57BL/10ScCR strains, and these animals have provided a valuable model for determining host immune mechanisms which are regulated by the Lps gene.

2.1 Lps^d Gene Regulation of IgA Immune Responses

Immunoglobulin A (IgA) is the major isotype of Ig found in external secretions. Since greater than 95% of all infectious diseases occur on, or begin with penetration of pathogens through the mucosal surfaces, it is important to precisely evaluate the influence of gut LPS on the lymphoid tissues which are major IgA-inductive sites. The first major organized lymphoid tissue encountered by endogenous LPS is the gut-associated lymphoreticular tissue (GALT), which mainly consists of the Peyer's patches (PP). The PP are distinct follicles located along the gastrointestinal wall of most experimental animals, including mice. The PP represent the major inductive site for IgA responses to ingested antigens and pathogenic microorganisms in the gastrointestinal tract in mammals.

The dome region of the PP is covered by an epithelium which contains a unique cell type termed an M cell (OWEN and JONES 1974) or follicular-associated epithelial (FAE) cell (BOCKMAN and COOPER 1972). The M cell is actively pinocytotic and phagocytic and samples both soluble and particulate antigens (e.g., viruses and even whole bacteria) which are present in the lumen of the gut. The M cell can then present antigen to underlying lymphoreticular cells, which leads to the sensitization of lymphoid cells present in distinct T- and B-cell zones in the PP.

Recent studies have allowed a complete characterization of the lymphoreticular cell types found in the murine PP (KIYONO et al. 1982a). Approximately 40% of PP cells are B-lymphocytes, a high percentage (12%–16%) of which bear surface IgA and are committed to IgA responses. Regulatory T cells also represent approximately 40% of the PP cell population, and include 40%–50% Lyt-1$^+$ inducer and helper cells, and 15%–20% Lyt-2$^+$ cytotoxic and suppressor cells. A significant percentage of PP T cells bear Fc receptors for IgA and are important in regulation of IgA isotype responses. The PP also contains accessory cells including macrophages (5%–9%) and functional dendritic cells (SPALDING et al. 1983). Considerable emphasis is currently being placed on the identification of the role of each cell type in the induction and regulation of IgA isotype responses.

Although PP are a major site for induction of IgA responses, B-cell differentiation and antibody synthesis does not occur in this tissue following immunization with gut lumenal antigens. Instead, antigen-sensitized IgA-committed B cells and T cells leave PP via efferent lymphatics, and pass through mesenteric lymph nodes and the thoracic duct lymphatics to enter the bloodstream. From the blood, IgA-committed B cells migrate to and selectively localize in distant mucosal tissues, e.g., mammary, salivary, and lacrimal glands, and lamina propria regions of the gastrointestinal and upper respiratory tracts. The B cells in these sites then differentiate into plasma cells, presumably via interactions with regulatory T cells. The polymeric IgA antibody produced by these plasma cells is specific for the antigen first encountered in GALT and is selectively transported through epithelial cells and into external secretions. The ability of inductive events in GALT to give rise to simultaneous IgA responses in several mucosal sites has led to the concept of a common mucosal immune system (CMIS) which can provide the host with an efficient immune mechanism for protection of mucous membranes against potentially detrimental antigens and pathogens. A more detailed consideration of the induction and regulation of IgA responses and the CMIS has been the subject of a recent book (McGhee and Mestecky 1983) and several reviews (McGhee et al. 1984; Tomasi 1983).

Since LPS is ubiquitous and a major macromolecular constituent of the gut, it has been difficult to precisely determine the contribution of endogenous LPS to the maturation and function of cells of the immune system. The now classic work of Dubos and Schaedler (1960) suggested that the indigenous gram-negative microflora of the gastrointestinal tract has a marked effect on the hosts' susceptibility to infection. Evidence has been presented that specific pathogen-free and germ-free mice are susceptible to infection (Dubos and Schaedler 1960), but resistant to lethal effects of endotoxin (Jensen et al. 1963; Schaedler and Dubos 1961, 1962), while mice which possess a gram-negative gut microflora are more susceptible to the lethal effects of endotoxin (Jensen et al. 1963; Schaedler and Dubos 1962; Kiyono et al. 1980b). These results suggest that endogenous LPS in the gastrointestinal tract influences host susceptibility to gram-negative infection, perhaps via an interaction with GALT lymphoreticular cells and the subsequent induction of innate host immunity to infection.

Our studies have been directed toward the determination of the role of gut LPS in modulating the ability of GALT cells to respond to TD antigens. We have compared immune responses to orally administered TD antigens in mice hyporesponsive to LPS, e.g., the C3H/HeJ mouse strain, with those of identically treated, fully syngeneic, LPS-responsive (Lps^n/Lps^n) C3H/HeN mice. C3H/HeJ mice elicit higher IgA responses, including splenic plaque-forming cell (PFC) and serum antibody responses, to orally administered heterologous erythrocytes than identically treated C3H/HeN mice (Babb and McGhee 1980). Elevated IgA anti-erythrocyte responses in C3H/HeJ mice are largely due to the induction of T helper (Th) cell activity in GALT (and in spleen) with minimal induction of T suppressor (Ts) cell activity (Kiyono et al. 1980a). These results suggest that the Lps^d allele influences the induction of IgA responses to orally administered TD antigens. As will be discussed in detail below, sustained oral immunization of Lps^n/Lps^n mice with sheep erythrocytes (SRBCs) resulted in elevated Ts cell activity and subsequent systemic unresponsiveness (oral tolerance).

Additional evidence that elevated IgA responses in mice were related to Lps^d expression was provided by studies in which mice from four different LPS-responsive and two LPS-hyporesponsive mouse strains were orally primed with SRBCs and subsequently orally immunized with trinitrophenyl (TNP)-haptenated SRBCs. The four Lps^n/Lps^n mouse strains gave low but similar splenic IgA anti-TNP PFC responses, while both Lps^d/Lps^d strains gave high

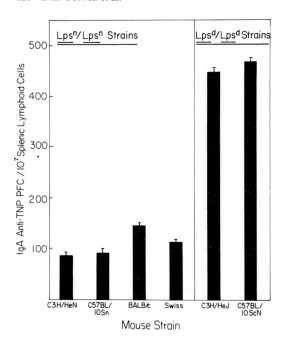

Fig. 2. Influence of the Lps^d gene on IgA immune responses. Splenic IgA anti-TNP responses in various LPS-responsive (Lps^n/Lps^n) and -hyporesponsive (Lps^d/Lps^d) mouse strains, carrier primed (SRBCs) and immunized (TNP-SRBCs) by gastric intubation

IgA anti-TNP PFC responses (Fig. 2). This finding suggested an association between the ability of mice to respond to LPS and the magnitude of IgA responses induced by orally administered TD antigens.

Formal proof that IgA responses are regulated by the Lps^d gene was provided by studies with offspring from genetic crosses with C3H/HeN and C3H/HeJ mice, i.e., (C3H/HeJ × C3H/HeN) F1, F2 and backcross mice (MICHALEK et al. 1980). In this investigation, mice were orally primed with SRBCs followed by oral immunization with TNP-SRBCs and splenic IgA anti-TNP PFC responses and mitogenic responses to LPS were determined. Hybrid (C3H/HeJ × C3H/HeN) F1 mice exhibited intermediate IgA anti-TNP PFC and LPS mitogenic responses when compared with those of the parental strains (Fig. 3). When F2 mice were examined, approximately one-half exhibited an intermediate response pattern similar to that observed with F1 hybrids. Approximately 23% (13 out of 57) of the F2 mice gave high LPS mitogenic and low IgA anti-TNP PFC responses, while 19% (11 out of 57) exhibited a response pattern characteristic of C3H/HeJ mice (Fig. 3). Definitive evidence that the level of IgA responses to orally administered TD antigen correlated with responsiveness to LPS was obtained using backcross mice. Offspring from crosses between F1 and C3H/HeN mice segregated into two distinct response groups; one-half of the mice gave intermediate IgA immune and LPS mitogenic responses (characteristic of F1 mice), while the remainder of the mice exhibited a response pattern similar to that obtained with C3H/HeN mice (Fig. 3). Analysis of (F1 × C3H/HeJ) backcross mice revealed a response pattern reflective of either F1 or C3H/HeJ mice. These results demonstrated that IgA responses to orally administered TD antigens are regulated by the Lps gene (MICHALEK et al. 1980).

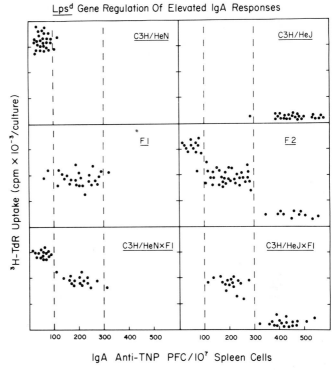

Fig. 3. *Lps*^d gene regulation of IgA responses to orally administered TD antigen (see Fig. 2 legend). Each point is the mean of quadruplicate spleen cell cultures for LPS mitogenic responses and triplicate determinations for splenic IgA anti-TNP PFC responses per animal

In other studies, we have shown that oral immunization of C3H/HeJ mice with other TD antigens, including horse, chicken or human erythrocytes, or gram-positive bacteria (*Streptococcus mutans*), results in significant IgA responses as assessed by splenic PFC and serum and salivary antibody responses (BABB and McGHEE 1980; KIYONO et al. 1982c; and unpublished observations). In recent studies (JIRILLO et al. 1984a), we have assessed the ability of C3H/HeJ mice to mount IgA responses to the three major regions of the LPS molecule when appropriate antigen was administered by the oral route. For these studies, we used either smooth or rough strains of *Salmonella*, i.e., *S. typhimurium* LT-2, whose LPS is composed of an outer O-polysaccharide chain, a central core region, and lipid A; *S. minnesota* R345 (Rb), whose LPS is missing the entire O-polysaccharide chain; and *S. minnesota* R595 (Re), whose LPS consists only of lipid A and three ketodeoxyoctonate (KDO) sugars. Groups of C3H/HeN and C3H/HeJ mice were given whole, killed *Salmonella* bacteria by gastric intubation for three consecutive days/week for 2 weeks and then were boosted intravenously 1 week later with homologous, purified LPS. C3H/HeJ (*Lps*^d/*Lps*^d) mice gave higher splenic IgA anti-LPS PFC responses to the three different *Salmonella* LPS preparations than did similarly treated C3H/HeN mice. The splenic PFC response in mice given smooth *Salmonella* was primarily directed

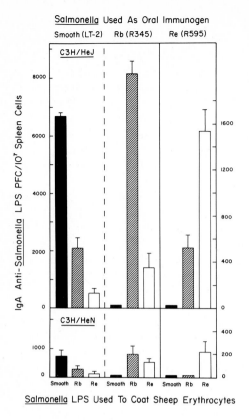

Fig. 4. Induction of IgA anti-*Salmonella* LPS responses in *Lps*^{d}/*Lps*^{d} mice given *Salmonella* whole cell antigen by the oral route. Groups of mice were given whole *Salmonella* cells orally for three consecutive days per week for 2 weeks. Following an intravenous boost of mice with homologous LPS, splenic IgA anti-LPS PFC responses were determined

toward the homologous LPS, although some reactivity against Rb and Re LPS was noted (Fig. 4). Similarly, mice orally immunized with either Rb or Re *Salmonella* gave responses which were primarily directed toward the R core or lipid A, respectively, and no response was detected when LPS from smooth *Salmonella* was used to coat the indicator erythrocytes. In additional studies, saliva from orally immunized C3H/HeJ mice was shown to contain higher levels of IgA anti-LPS antibodies than saliva from similarly treated C3H/HeN mice. These results indicate that although C3H/HeJ mice are hyporesponsive to LPS given systemically, they are capable of making IgA responses to the three major regions of LPS when killed gram-negative bacteria are administered orally, thus suggesting that IgA responses to LPS occur independently of lipid A mitogenic signals.

2.2 Influence of Gut LPS on Oral Tolerance Induction

As previously discussed, prolonged oral administration of TD antigen to mice results in the induction of oral tolerance in *Lps*^{n}/*Lps*^{n} mice but not in *Lps*^{d}/*Lps*^{d} animals. One possible explanation for this finding is that endogenous gut LPS induces the production of increased numbers of Ts cell precursors and subsequent elevation of Ts cell activity in GALT of Lps^n/Lps^n mice (reviewed by McGHEE et al. 1984). Alternatively, endogenous gut LPS may alter the contrasuppressor T-cell circuit, which could also account for the enhanced Ts cell activity in GALT. In this regard, it has been shown that the contrasuppressor T-cell circuit is an important form of immunoregulation in murine PP (GREEN et al. 1982). Thus, LPS could influence host immunity, since the induction of Ts cells and their migration to peripheral lymphoid tissue, such as the spleen, would significantly diminish systemic immune responses. It is now established that intragastric administration of large doses of soluble protein or prolonged feeding of heterologous erythrocytes results in the inability of the host to respond to systemically administered antigen, a condition termed oral tolerance (TOMASI 1980). Although oral tolerance may be mediated by several different pathways, the evidence demonstrating the induction of Ts cells in GALT and their subsequent migration to peripheral tissue strongly supports a role for Ts cells in mediating systemic unresponsiveness. The induction of Ts cell activity in GALT does not necessarily lead to suppression of IgA responses, since it has been shown that oral tolerance occurs with concomitant expression of secretory IgA responses (CHALLACOMBE and TOMASI 1980).

If systemic unresponsiveness is mediated by Ts cells from GALT in Lps^n/Lps^n mice, one may postulate that GALT of C3H/HeJ mice possesses fewer Ts-cell precursors than Lps^n/Lps^n mice or possesses an enhanced contrasuppressor circuit. In either case, Lps^d/Lps^d mice offer a model for separation of T-cell-regulated IgA responses and Ts-cell-mediated oral tolerance. To test this, groups of Lps^n/Lps^n and Lps^d/Lps^d mice were given SRBCs daily by gastric intubation for two consecutive weeks and then were tested for their ability to respond to this antigen administered systemically. All Lps^n/Lps^n mouse strains were unresponsive to SRBCs, while the two Lps^d/Lps^d mouse strains gave anamnestic responses, especially of the IgA isotype (Fig. 5). These results suggest that Lps^d/Lps^d mice are insensitive to oral tolerance induction with SRBCs and that continued oral administration of antigen actually primes these mice for secondary responses (KIYONO et al. 1982b).

Oral tolerance induction in Lps^n/Lps^n mice and secondary responses to SRBCs in C3H/HeJ mice were mediated by regulatory T cells in GALT (KIYONO et al. 1982b; MICHALEK et al. 1982). When T-cell-enriched fractions from the spleen and PP of mice given SRBC daily by gastric intubation for 2 weeks were assessed for effector function, C3H/HeJ T cells principally showed helper activity and were of the $Lyt-1^+$ phenotype, whereas C3H/HeN T cells primarily exhibited Ts cell activity and were of the $Lyt-2^+$ phenotype. The demonstration that treatment of C3H/HeN PP or spleen cells with anti-Lyt-2.1 and complement abrogated suppression provided additional evidence that the Ts cells originated in PP and mediated oral tolerance in systemic lymphoid tissues.

In other studies, we have noted that (C3H/HeJ × C3H/HeN) F1 mice given SRBCs daily by gastric intubation for 2 weeks gave lower splenic immune responses to systemically administered SRBCs than C3H/HeJ mice (MICHALEK et al. 1982). This observation suggests that while the continued oral administration of SRBCs to F1 mice induces both Th and Ts cell activity, there is a

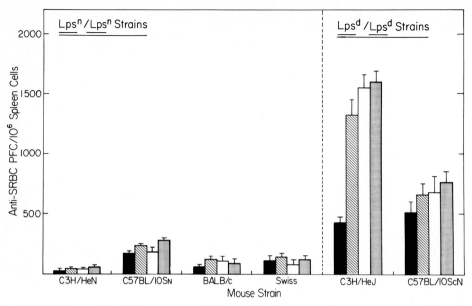

Fig. 5. Influence of the Lps^d gene on oral tolerance induction. Groups of Lps^n/Lps^n and Lps^d/Lps^d mice were given SRBCs daily for 14 consecutive days by gastric intubation, and following intravenous challenge with SRBCs, the splenic IgM (■), IgG$_1$ (▨), IgG$_2$ (□), and IgA (▥) anti-SRBC PFC responses were assessed

predominance of Th cell activity which could account for the observed systemic responses to SRBCs (MICHALEK et al. 1982).

The results summarized above provide indirect evidence that endogenous gut LPS augments a Ts-cell pathway in GALT of Lps^n/Lps^n mice. More direct proof for the role of LPS in immune suppression was obtained by studies in mice which had received minimal exposure to LPS but were fully responsive to this macromolecule. Systemic administration of SRBCs to germ-free C3H/ HeN, BALB/c, or Swiss mice which had been previously given SRBCs daily by gastric intubation for 2 weeks resulted in secondary immune responses. In contrast, identically treated conventionally reared mice of these three strains were unresponsive (WANNEMUEHLER et al. 1982). Of interest was the demonstration that prior intragastric administration of LPS to germ-free mice rendered these animals susceptible to oral tolerance induction by SRBCs (Fig. 6). An oral dose of as little as 10 µg LPS significantly reduced systemic responses in germ-free mice, whereas a dose of 100 µg or more LPS given orally to these animals rendered them completely sensitive to oral tolerance induction. Germ-free mice given antigen alone predominantly exhibited Th-cell activity in GALT and spleen, whereas germ-free mice which were treated with LPS prior to oral administration of antigen had significant Ts cell activity in these tissues. These results provide direct evidence that endogenous gut LPS induces the production of precursor Ts cells in GALT and that this cell type mediates oral tolerance induction.

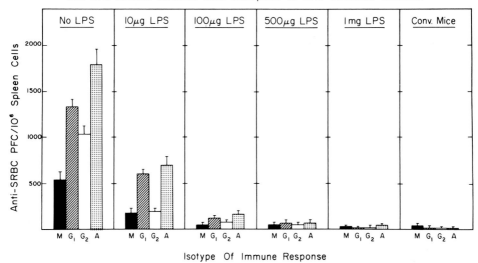

Fig. 6. Direct evidence for gut LPS influence on oral tolerance induction. Groups of germ-free mice were given various doses of LPS prior to prolonged oral administration of SRBCs, and splenic anti-SRBC PFCs were assessed (see Fig. 5 legend for details)

   This LPS effect would clearly be beneficial to the host, since endogenous LPS would render GALT cells capable of tolerance induction in response to large influxes of ingested environmental antigens. In this manner, endogenous gut LPS could prevent continuous systemic responses to environmental antigens which escape the gut and reach the circulation, and prevent the deleterious effects, including inflammation and target organ damage, which are associated with the formation and deposition of immune complexes. In this respect, the effect of LPS on Ts-cell precursors in GALT would provide the host with a second major defense against environmental antigens, the first, of course, being immune exclusion of antigens by secretory IgA antibodies at the mucosal surface itself.

2.3 Lymphoreticular Cell Responses to *Bacteroides* LPS

All of the studies summarized above on LPS effects on GALT and responses to TD antigen were performed with LPS derived from *Enterobacteriaceae*, i.e., *E. coli* or *Salmonella* species, which comprise less than 1% of the gram-negative microflora of the gastrointestinal tract. Greater than 99% of the gut flora is represented by *Bacteroides* species and recent studies have suggested that *Bacteroides fragilis* LPS triggers B-cell responses in Lps^d/Lps^d mice (JOINER et al. 1982). This finding suggests that the *Lps* gene only regulates responses to lipid A from *Enterobacteriaceae*. If this, indeed, were true, then only a minor component of the gut microflora would account for our results summarized above, which

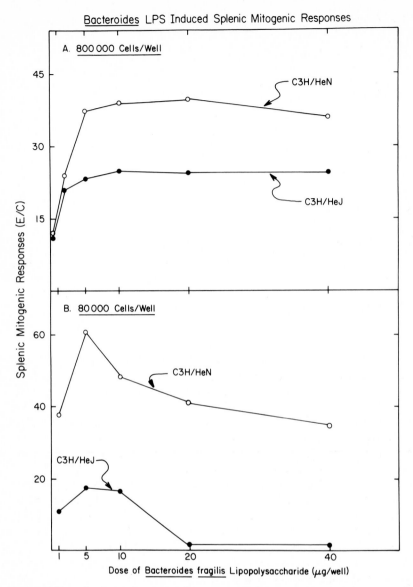

Fig. 7. Effect of spleen cell numbers from C3H/HeN (Lps^n/Lps^n) and C3H/HeJ (Lps^d/Lps^d) mice on mitogenic responses to *B. fragilis* LPS

suggest that gut LPS and the *Lps* gene regulate IgA responses and oral tolerance to orally administered, TD antigens.

Therefore, we have reexamined whether responses to *Bacteroides* LPS are regulated by the *Lps* gene, by testing the ability of classical LPS-responsive C3H/HeN and LPS-hyporesponsive C3H/HeJ mice to respond to a phenol-water extract of *B. fragilis* ATCC 25285 LPS (B-LPS). When B-LPS was added

Table 2. Mitogenic and polyclonal IgM responses to *Bacteroides* LPS of purified splenic B cells from C3H/HeN, (C3H/HeN × C3H/HeJ)F1, and C3H/HeJ mice

	Source of splenic B cells		
	C3H/HeN	F1	C3H/HeJ
Mitogenic responses[a] B-LPS			
20 µg	29.8	10.7	1.8
10 µg	22.0	7.4	1.3
Polyclonal IgM synthesis[b] B-LPS			
50 µg	23620	7170	375
10 µg	28087	6675	735

[a] Purified splenic B cells (5×10^5/well) were cultured with the indicated amount of B-LPS at 37° C in 5% CO_2 in air and pulsed with ^{3}H-TdR during the last 6 h of a 48-h incubation. Data are expressed as the stimulation ratio (E/C)
[b] Purified splenic B cells were cultured with B-LPS at 37° C in 5% CO_2 in air. Culture supernatants were harvested on day 7 and levels (nanograms/culture) of IgM were assessed by RIA

to either C3H/HeN or C3H/HeJ spleen cell cultures at either high or low cell density, C3H/HeJ spleen cells responded to all B-LPS doses tested at high cell concentration; however, the response was less than that seen in C3H/HeN spleen cell cultures (Fig. 7) (WANNEMUEHLER et al. 1984). On the other hand, when low concentrations of spleen cells were tested, C3H/HeJ cultures exhibited lower responses to all B-LPS doses than were observed in C3H/HeN cultures. This finding suggested that C3H/HeJ mice may be low responders to B-LPS just like to LPS from *Enterobacteriaceae* (WANNEMUEHLER et al. 1984).

The difference in B-LPS mitogenic responses was more marked when purified splenic B cells from these two mouse strains were tested. Purified splenic B cells from C3H/HeN mice gave good mitogenic responses and polyclonal IgM synthesis in response to B-LPS, while splenic B cells from Lps^d/Lps^d mice were unresponsive to B-LPS (Table 2). Splenic B cells from (C3H/HeN × C3H/HeJ) F1 mice gave intermediate mitogenic and polyclonal IgM response patterns. This suggests that B-cell responses to B-LPS are regulated by the Lps^d gene (WANNEMUEHLER et al. 1984).

In another series of studies, purified lipid A and carbohydrate were prepared from B-LPS by mild acid hydrolysis and were tested for mitogenic activity in either C3H/HeN or C3H/HeJ B cell or whole spleen cell cultures (Table 3). *Bacteroides* lipid A induced mitogenic responses in cultures from C3H/HeN mice and failed to trigger C3H/HeJ cell cultures. The carbohydrate moiety of B-LPS failed to induce responses in B-cell cultures from either strain; however, good mitogenic responses were induced in whole spleen cell cultures from these animals (Table 3) (WILLIAMSON et al. 1984). In other experiments, we have shown that macrophages are required for the carbohydrate-induced re-

Table 3. Mitogenic responses of purified splenic B cells and whole spleen cells from C3H/HeN and C3H/HeJ mice to *Bacteroides* LPS and the lipid A and polysaccharide moieties[a]

LPS preparation	Splenic B cells		Whole spleen cells	
	C3H/HeN	C3H/HeJ	C3H/HeN	C3H/HeJ
B. fragilis				
LPS(Ph)	18.0	1.5	18.8	2.6
Lipid A	12.3	1.1	8.2	1.4
Polysaccharide	1.4	1.3	11.1	12.9
E. coli K235				
LPS(Ph)	21.2	1.4	29.8	1.7
Lipid A	14.9	1.5	24.9	2.0
Polysaccharide	2.5	1.0	1.7	1.8

[a] See Table 2, footnote [a]. LPS preparations were added at 10 µg/culture. Whole spleen cells were cultured at 5×10^5/well

sponse and splenic macrophages from either C3H/HeN or C3H/HeJ mice restore C3H/HeJ B-cell responses to the polysaccharide. Polysaccharide prepared from *E. coli* K235 failed to induce responses in any cell cultures, clearly indicating that the polysaccharide component of *Bacteroides* LPS is biologically active (WILLIAMSON et al. 1984).

From these experiments, we conclude that the lipid A moiety of B-LPS is biologically active and induces mitogenic and polyclonal responses which are regulated by the *Lps* gene. The apparent B-cell mitogenic responses seen in C3H/HeJ spleen cell cultures are due to a biologically active polysaccharide moiety which requires macrophages for triggering B-cell responses. Thus, the *Lps* gene regulates *Bacteroides* lipid A responses and the lipid A effects of endogenous gut LPS on GALT cells should be similar to those induced by *Enterobacteriaceae* LPS.

3 Influence of the *Lps* Gene on Susceptibility to *Salmonella* Infection

Investigators have long been interested in the regulation of host resistance to infection, and a great volume of literature on this subject has accumulated. Early observations of differences in susceptibility of various animal species to bacterial, viral, protozoan, and helminthic parasites (reviewed by GOWEN 1948) led to the idea that resistance to infection may be genetically conferred. The studies of Hill, Gowen, Schott, Webster, and others (reviewed by GOWEN 1948) in which inbred lines of mice of progressively increasing resistance were developed by selective breeding of the survivors of challenge with pathogenic bacteria or viruses demonstrated that inheritance strongly influenced susceptibility to naturally occurring or experimentally induced infection. WEBSTER (1933, 1937) further analyzed the genetic origin of resistance to infection by examining the inheritance of resistance in F1 progeny from crosses between innately resistant (15% mortality) and innately susceptible (85%–95% mortality) mouse strains. Approximately 18% of the (resistant × susceptible) F1 progeny succumbed to infection with *Salmonella enteritidis* or St. Louis encephalitis virus regardless of the sex of the animal. In addition, susceptibility and resistance segregated in backcross progeny, suggesting that resistance was inherited as a single-factor Mendelian dominant, non-sex-linked trait (WEBSTER 1933, 1937).

Numerous later studies have reported marked differences in the susceptibility of various commonly used inbred mouse strains to infection with gram-negative bacteria, focussing in particular on resistance and susceptibility of mice to infection with virulent *Salmonella typhimurium*. As a result of these studies, inbred mouse strains have been designated as either innately resistant or innately susceptible to *Salmonella* infection, or, in the case of DBA/2J mice challenged intraperitoneally with *S. typhimurium*, of intermediate susceptibility, based upon the numbers of bacteria required to kill 50% of the mice (LD_{50}) of a given strain. Thus, BSVS (WEBSTER 1933), BALB/c, B10.D2, C57BL/6, C3H/HeJ, DBA/1 (ROBSON and VAS 1972), and CBA/N mice (O'BRIEN et al. 1979) have been classified as *Salmonella* susceptible (subcutaneous, intravenous, or intraperitoneal $LD_{50} < 20$ bacteria; O'BRIEN et al. 1980a), while BRVR (WEBSTER 1933), A/J, C3H/He (ROBSON and VAS 1972), CBA, DBA/2 (challenged subcutaneously; PLANT and GLYNN 1976), CBA/Ca (O'BRIEN et al. 1979), C3H/Bi, C3H/HeN, C3H/St (O'BRIEN et al. 1980b), A/HeN, C57/L (challenged subcutaneously), and SWR/J (O'BRIEN et al. 1980a) mice have been grouped as *Salmonella* resistant (subcutaneous, intravenous, or intraperitoneal $LD_{50} > 1 \times 10^3$ bacteria; reviewed by O'BRIEN et al. 1981a). It is important to note at this point that the susceptibility or resistance of a given mouse strain to experimental *Salmonella* infection is highly dependent not only upon the genetic constitution of the mouse strain, but also upon the virulence of the *Salmonella* strain used and the route of challenge employed.

3.1 Genetic Control of Host Susceptibility to *Salmonella* Infection

The results of WEBSTER's breeding experiments (1933, 1937) and the striking contrast between LD_{50}s of susceptible and resistant inbred mouse strains (see above) suggested that resistance to *Salmonella* infection in mice was regulated by either a single dominant gene or a cluster of closely linked genes. GOWEN (1960), however, observed a broad spectrum of degrees of resistance of ten different mouse strains to *S. typhimurium* infection, and interpreted this finding as an indication that resistance to *Salmonella* in the mouse was under polygenic control. Similarly, Hormaeche (reviewed by HORMAECHE et al. 1980) reported that the genetic control of resistance to *Salmonella* infection in mice was complex, with the early in vivo net growth rate of *S. typhimurium* being regulated by a single autosomal gene or cluster of genes, and later phases of infection being controlled by multiple separate genes.

In 1976, PLANT and GLYNN described a gene which governed the resistance of CBA (resistant) and BALB/c (susceptible) mice to subcutaneous challenge with *S. typhimurium*, which exhibited simple Mendelian inheritance in (BALB/c × CBA) F1, F2, and backcross generations, and which was not linked to the H-2 gene complex. This gene, designated *Ity* (for "immunity to typhimurium"), has been reported to consist of a dominant allele *Ity*^{r}, carried by resistant mice, and a recessive allele *Ity*^{s}, for which susceptible mice are homozygous, and has been mapped to a position on chromosome 1 closely linked to the *ln* locus which encodes leaden coat color in mice (PLANT and GLYNN 1979). Although the mechanism by which *Ity* controls resistance to *Salmonella* infection is unknown, a macrophage deficit or a defect in the phagocytic or microbicidal activities of splenic and hepatic macrophages has been suggested to be responsible for the inability of *Ity*^{s}/*Ity*^{s} mice to control early net multiplication of *Salmonella* (reviewed by O'BRIEN et al. 1981a). In this regard, LISSNER et al. (1983) have recently demonstrated that the *Ity* gene does not regulate phagocytosis of *S. typhimurium* in vivo, but rather the efficiency of killing of intracellular *Salmonella* by macrophages.

In 1979, O'BRIEN et al. reported that B-cell-defective CBA/N mice were highly susceptible ($LD_{50} = 20$ bacteria) to intraperitoneal challenge with virulent *S. typhimurium*, while immunologically normal CBA/CaHN mice were highly resistant ($LD_{50} = 2 \times 10^4$ bacteria). This finding suggested that the greater susceptibility of CBA/N mice might be associated with the expression of the X-linked *xid* gene in this mouse strain. The finding that F1 male progeny from crosses between immune-defective CBA/N mice and immunologically normal mice had LD_{50}s of ≤ 20 bacteria, while the corresponding F1 female mice had LD_{50}s of $\geq 1 \times 10^4$ organisms, and the finding that expression of the *xid* gene (as assessed by low levels of IgM antibody in the serum) closely correlated with the mortality of *Salmonella*-infected (CBA/N × immunologically normal) F1 male, F2, and backcross mice provided further support for the supposition that susceptibility of CBA/N mice to *Salmonella* infection was influenced by the *xid* gene (O'BRIEN et al. 1979). In addition, O'BRIEN and coworkers (1979) suggested that CBA/N mice carry a resistant allele of the autosomal gene (*Ity*) described by PLANT and GLYNN (1979) and that expression of the resistant allele in CBA/N mice is prevented by expression of the *xid* gene, since they found that (*Salmonella*-susceptible CBA/N × *Salmonella*-susceptible BALB/c) F1 female mice were resistant to intravenous challenge with *S. typhimurium*. The demonstration that (CBA/N × DBA/2N) F1 male mice developed a poor IgG anti-*S. typhimurium* antibody response following immunization with killed bacteria and that the resistance of F1 male mice to infection with virulent *S. typhimurium* could be significantly enhanced either by the passive transfer of immune sera from F1 female littermates, or by reconstitution of lethally irradiated F1 male mice with bone marrow cells from F1 female mice or (DBA/2N × CBA/N) F1 male mice, suggested that the *xid* gene conferred *Salmonella* susceptibility upon CBA/N mice via a defective humoral immune response to the invading microorganism (O'BRIEN et al. 1981b). These findings emphasize the importance of humoral as well as cellular immune elements in defense of the host against *Salmonella* infection (for further discussion see O'Brien, this volume).

3.2 Evidence for *Lps* Gene Regulation of Susceptibility to *Salmonella* Infection

When the degrees of susceptibility of nine different inbred mouse strains and one F1 hybrid strain to intraperitoneal challenge with virulent *S. typhimurium* F1826 were compared with the levels of mitogenic responsiveness of spleen cells from the respective mouse strains stimulated in vitro with LPS extracted from homologous *Salmonella* by the phenol-water method, it was found that the mouse strains which were most resistant to bacterial challenge [i.e., C3Hf, CBA/H, and (C57BL/6 × C3Hf)F1] had the highest levels of tritiated-thymidine incorporation in response to LPS stimulation, while the strain which was least resistant to challenge (C3H/HeJ) had the lowest level (VON JENEY et al. 1977). This finding suggested that there might be a correlation between the ability of spleen cells from a mouse to be mitogenically stimulated by LPS and the degree of resistance to *S. typhimurium* infection exhibited by that mouse strain.

To determine whether this apparent correlation between LPS sensitivity and *Salmonella* susceptibility in mice could be extended to the genetic basis for these two phenomena, O'BRIEN and coworkers (1980a, b) examined the relationship between expression of the Lps^d allele and susceptibility of mice to infection with *S. typhimurium* TML. As discussed in Sect. 3 above, these investigators demonstrated that C3H/HeJ mice (Lps^d/Lps^d) were highly susceptible to parenteral challenge with *S. typhimurium* ($LD_{50} < 2$), while syngeneic C3H/HeN (Lps^n/Lps^n) mice were highly resistant ($LD_{50} = 2 \times 10^3$; O'BRIEN et al. 1980a, b). That the *Ity* gene was not responsible for the susceptibility of C3H/HeJ mice was demonstrated by the finding that F1 hybrid progeny from C3H/HeJ and C57BL/6J (Ity^s/Ity^s) parents were resistant ($LD_{50} \geq 8 \times 10^3$) to infection with *S. typhimurium*, thus suggesting that the genes which controlled *S. typhimurium* susceptibility in C3H/HeJ and C57BL/6J mice were distinct and that C3H/HeJ mice carried a gene, presumably Ity^r, which complemented Ity^s in the F1 hybrid animals (O'BRIEN et al. 1980). In addition, since both male and female (C3H/HeJ × C3H/HeN) F1 mice were resistant ($LD_{50} \geq 1 \times 10^4$) to *S. typhimurium* infection, it was concluded that susceptibility of C3H/HeJ mice was governed by a non-sex-linked, recessive allele of a resistance gene and was, therefore, controlled by a gene other than *xid* (O'BRIEN et al. 1980). The finding that 51% of the backcross progeny derived from C3H/HeJ and (C3H/HeJ × C57BL/6J)F1 parents succumbed to *S. typhimurium* infection, while 44% survived was consistent with the proportions of susceptible and resistant offspring expected if *S. typhimurium* susceptibility were inherited as a trait regulated by a single gene locus with dominant and recessive alleles (O'BRIEN et al. 1980). Moreover, linkage analyses performed in backcross progeny revealed a close correlation between the phenotypic expression of the $Mup-1^a$ allele and susceptibility to *S. typhimurium*, and between susceptibility to *S. typhimurium* and hyporesponsiveness to LPS, thus indicating that the susceptibility of C3H/HeJ mice to *S. typhimurium* infection was determined by a gene on chromosome 4 which was either identical to or closely linked to the *Lps* gene (O'BRIEN et al. 1980).

In contrast, there is no clear evidence to suggest that resistance or susceptibility to *S. typhimurium* infection in other mouse strains of the C3H lineage is controlled by the *Lps* gene. In fact, although O'BRIEN and coworkers (1980b) reported an inverse relationship between the degree of susceptibility of C3H/HeJ, C3H/HeN, C3H/St, and C3H/Bi mice to infection with *S. typhimurium* and the capacity of lymphocytes from these mouse strains to be mitogenically stimulated by phenol-water extracted LPS from *Escherichia coli* K235, EISENSTEIN et al. (1982) found no consistent correlation between *S. typhimurium* susceptibility and LPS responsiveness of C3HeB/FeJ, C3H/HenCr1BR, C3H/BiMA, C3H/HeDub, and C3H/HeTex mice. The latter investigators, for example, demonstrated that while C3H/HeJ mice were LPS hyporesponsive and highly susceptible to infection with *S. typhimurium* W118-2 (intraperitoneal LD_{50} < seven organisms), and C3H/HeNCr1BR mice were LPS responsive and highly resistant to *S. typhimurium* infection (intraperitoneal $LD_{50} = 1.2 \times 10^3$ organisms), C3HeB/FeJ mice were LPS responsive and hypersusceptible to *S. typhimurium* (intraperitoneal LD_{50} < two bacteria; EISENSTEIN et al. 1982). Furthermore, the rapid multiplication of intravenously injected *S. typhimurium* detected in the livers and spleens of C3HeB/FeJ mice during the first 5 days post challenge resembled, but was not identical to, the early net growth rate of *S. typhimurium* in the tissues of infected C3H/HeJ mice (EISENSTEIN et al. 1982), and lended support to the previous reports that *S. typhimurium* proliferation followed an early "fast" growth rate in the liver and spleen of susceptible mouse strains and a "slow" growth rate early in infection in resistant mice (reviewed by O'BRIEN et al. 1981a).

More recently, O'BRIEN and ROSENSTREICH (1983) have reported that the degrees of *S. typhimurium* multiplication in the spleens of C3H/HeJ and C3HeB/FeJ mice by day 10 following subcutaneous challenge were similar, but that the levels of *S. typhimurium* in the spleens of (C3HeB/FeJ × C57BL/6J)F1, (C3HeB/FeJ × C3H/HeJ)F1, and (C3HeB/FeJ × C57L/J)F1 male and female mice 10 days post challenge were significantly lower than the levels of *S. typhimurium* in the spleens of the respective parental strains. These investigators interpreted their results as an indication that the susceptibility of C3HeB/FeJ mice to infection with *S. typhimurium* was regulated by a gene other than *xid* which was distinct from the *Lps* gene, *Ity* gene, and as yet unidentified *Salmonella* susceptibility gene, of C3H/HeJ (Lps^d/Lps^d), C57BL/6J (Ity^s/Ity^s), and C57L/J mice, respectively. In addition, they concluded that gene complementation accounted for the increased ability of F1 hybrid progeny to control proliferation of *S. typhimurium* in the spleen (O'BRIEN and ROSENSTREICH 1983). O'BRIEN and ROSENSTREICH (1983) further speculated that responsiveness to LPS and susceptibility to *Salmonella* infection in inbred mouse strains may be controlled by a pair of closely linked genes, and that the simultaneous expression of LPS responsiveness and *S. typhimurium* susceptibility in C3HeB/FeJ mice could result from the existence of a mutation in the allele responsible for *S. typhimurium* resistance but not in the allele governing LPS responsiveness (O'BRIEN and ROSENSTREICH 1983). Thus, although susceptibility to *S. typhimurium* infection appears to be regulated by the Lps^d allele in C3H/HeJ mice, one must consider the concurrent expression of additional genes which control resistance to *S. typhimurium* when one analyzes the influence of the *Lps* gene on immunity to *S. typhimurium* infection in either C3H/HeJ or other C3H-derived mouse strains.

3.3 Proposed Mechanisms by Which the *Lps* Gene Controls Susceptibility to *Salmonella* Infection

The mechanism by which the *Lps* gene controls susceptibility to *S. typhimurium* infection in C3H/HeJ mice is unknown, but it is undoubtedly complex. As outlined in Sect. 1, the defective *Lps*^{d} allele carried by C3H/HeJ mice is responsible for numerous abnormal responses to LPS (reviewed by SCIBIENSKI 1981). Furthermore, the defective LPS responsiveness of C3H/HeJ mice is exhibited by a number of different cell types, including B cells, T cells, fibroblasts, and macrophages (reviewed by VOGEL et al. 1981), which normally interact in nondefective mouse strains to provide resistance to *S. typhimurium* infection.

The importance of cell-mediated immunity in defense of the host against *S. typhimurium* infection has been well documented (reviewed by EISENSTEIN and SULTZER 1983). That the cell-mediated immune response of C3H/HeJ mice infected with *S. typhimurium* is defective was suggested by the finding that C3H/HeJ mice died early following parenteral challenge with this facultative intracellular parasite and that proliferation of the challenge organism in the livers and spleens of these mice progressed in an apparently uncontrolled fashion, presumably as a result of defective microbicidal activity of the hepatic and splenic macrophages (O'BRIEN et al. 1980b, 1982). When bone marrow reconstitution studies were performed to determine the nature of the cells responsible for the hypersusceptibility of C3H/HeJ mice to *S. typhimurium* infection, it was found that irradiated C3H/HeJ mice which received C3H/HeN bone marrow cells were significantly more resistant ($LD_{50} = 800$ organisms) to intraperitoneal challenge with *S. typhimurium* than were irradiated C3H/HeJ mice reconstituted with C3H/HeJ bone marrow cells ($LD_{50} =$ two bacteria) or irradiated C3H/HeN mice reconstituted with C3H/HeJ bone marrow cells (LD_{50} < three bacteria; O'BRIEN et al. 1982). These results indicated that resistance to *S. typhimurium* infection in mice was mediated by a radiosensitive bone-marrow-derived cell type(s) (i.e., B cell, T cell, macrophage, or natural killer cell), and that this cell type(s) was defective in C3H/HeJ mice.

One such cell type in C3H/HeJ mice for which numerous defects have been described is the macrophage. Macrophages from this mouse strain are hyporesponsive to the regulatory effects of highly purified LPS, including the stimulation of macrophage differentiation, and LPS-induced macrophage activation (reviewed by VOGEL et al. 1981). VOGEL and ROSENSTREICH (1979) suggested that the loss in ability of macrophages from C3H/HeJ but not C3H/HeN mice to bind and phagocytose antibody-coated sheep erythrocytes after 24 h in culture reflected a loss in Fc receptor expression by C3H/HeJ macrophages which resulted secondarily from the failure of these cells to maintain a differentiated state in vitro. Several investigators have concluded that the defect in activation of C3H/HeJ macrophages, as measured by glucose utilization, oxygen radical generation, and monokine production, was limited to the ability of these cells to respond to lipid A, since treatment of the macrophages with lymphokine-rich supernatant from concanavalin A (Con A) stimulated spleen cells, or infection of C3H/HeJ mice with the macrophage-activating agent Bacillus Calmette-Guérin (BCG) or the avirulent *S. typhimurium* strain SL3525 resulted in a dramatic increase in the sensitivity of C3H/HeJ macrophages to a number of lipid A-mediated effects (EISENSTEIN et al. 1984; reviewed by VOGEL et al. 1979). O'BRIEN and coworkers (1982), however, concluded that treatment of C3H/HeJ mice with viable BCG prior to challenge with *S. typhimurium* was insufficient to protect these animals against infection, since the extent of net multiplication of *S. typhimurium* in the spleens of BCG-primed C3H/HeJ mice resembled the rapid early net multiplication of *S. typhimurium* in the spleens of saline-primed C3H/HeJ mice. These investigators suggested that their results may reflect the failure of macrophages from BCG-primed C3H/HeJ mice to maintain an activated state long enough to control the invading *Salmonella*.

While the role of the macrophage in cell-mediated immunity to *Salmonella* infection in both LPS-responsive and LPS-hyporesponsive mouse strains has been clearly established (O'BRIEN et al. 1982; reviewed by EISENSTEIN and SULTZER 1983), the results of various studies suggest that other bone marrow-derived, radiation-sensitive cell types may also be involved in the resistance of LPS-responsive mice to *S. typhimurium* infection, and that defects in the biological activities of these cell types may contribute to the hypersusceptibility of C3H/HeJ mice to infection with *S. typhimurium*. It has been reported, for example, that the lipid A-induced differentiation of murine thymocytes in vitro and the LPS-enhanced production of interleukin 2 by Con A-stimulated T cells were defective in cells cultured from C3H/HeJ mice (reviewed by ROSENSTREICH and VOGEL 1981). In addition, NENCIONI and coworkers (1983) have recently reported that lymphocytes from gut-associated lymphoreticular tissue and from peripheral lymphoid organs of C3H/HeN mice possessed natural antibacterial activity against *S. typhimurium* in vitro, with the greatest level of activity being exhibited by lymphocytes from the Peyer's patches. The cell type responsible for the "spontaneous" antibacterial activity in the spleen and Peyer's patches was shown to be nonadherent, nonphagocytic, asialo GM1[+], Fc receptor[+], and Thy 1.2[−] (NENCIONI et al. 1983). The finding that the in vitro bactericidal activity of such lymphocytes from the spleen and Peyer's patches of C3H/HeN and CBA/J mice was significantly higher against *S. typhimurium* than that of lymphocytes from homologous tissues of C3H/HeJ, BALB/c, C57BL/10, and C57BL/6 mice suggested that these lymphocyte populations might contribute in vivo to the defense of the host by controlling replication of the *S. typhimurium* early in infection. In addition, this finding suggested, by extrapolation, that failure of these lymphocyte populations to control early bacterial multiplication, particularly at the gastrointestinal level where *Salmonella* normally invade the host, might contribute to the innate susceptibility of the latter mouse strains to *S. typhimurium* infection (TAGLIABUE et al. 1984).

In summary, studies of acquired immunity to infection in mice have demonstrated that a number of different cell types are involved either directly or indirectly in the defense of the host against the facultative intracellular parasite *S. typhimurium* (reviewed by EISENSTEIN and SULTZER 1983). The studies mentioned above suggest that effective immunity to *S. typhimurium* infection requires full functional capability of these different cells, since loss of function as a result of a genetically conferred defect in the ability of one or more of the cell types to respond to activation and/or differentiation stimuli such as LPS results in decreased resistance to infection. However, the exact mechanisms by which these different cell types contribute to host defense remain to be identified.

4 Summary

Bacterial endotoxin (or LPS) is ubiquitous in nature and has been shown to elicit a broad spectrum of beneficial as well as detrimental effects in a susceptible host. The availability of inbred mouse strains (i.e., C3H/HeJ, C57BL/10ScCR, and C57BL/10ScN) which are hyporesponsive to LPS has facilitated studies of the genetic and molecular bases of LPS-induced effects. Extensive genetic analyses performed during the 1970s with recombinant inbred mice and with F1, F2, and backcross progeny from LPS-hyporesponsive and LPS-responsive parents provided strong evidence that responsiveness to LPS in mice is regulated by a single gene, designated *Lps*, which is located on chromosome 4. LPS-responsive mice possess normal *Lps* alleles (Lps^n/Lps^n), while C3H/HeJ mice possess defective alleles (Lps^d/Lps^d) which render them refractory to all biological effects of LPS. Studies of the patterns of LPS responsiveness of F1, F2, and backcross progeny from crosses between Lps^n/Lps^n and C3H/HeJ (Lps^d/Lps^d) mice have shown that the alleles of the *Lps* gene are codominantly expressed.

Lipopolysaccharide is a potent modulator of host immune responses, and recent studies have shown that endogenous LPS from the microflora of the gut influences the induction of IgA immune responses and oral tolerance to orally administered thymic-dependent (TD) antigens. Evidence has been provided which suggests that endogenous LPS may exert this regulatory effect by inducing an expansion of the precursor Ts-cell population in gut-associated lymphoreticular tissue (GALT). These precursor Ts cells could then, following exposure to ingested TD antigen, differentiate into mature Ts cells which could regulate the induction of IgA immune responses. The migration of these Ts cells from GALT to systemic lymphoid tissues could in turn mediate systemic unresponsiveness (oral tolerance) to the antigen. One can speculate that this LPS effect would be beneficial to the host, since secretory IgA antibody is important in immune exclusion of antigens at mucosal surfaces, and oral tolerance induction is important in preventing systemic responses to antigens which could result in deleterious effects, such as immune complex formation and subsequent inflammation. LPS does not, however, induce these regulatory mechanisms in Lps^d/Lps^d mice, and oral administration of TD antigen to these animals leads to elevated IgA responses, and no oral tolerance. Because the secretory IgA response of Lps^d/Lps^d mice to ingested antigen is higher than that of Lps^n/Lps^n animals, the C3H/HeJ (Lps^d/Lps^d) mouse strain offers a unique model for assessing the role of secretory IgA antibodies in protection against infection by pathogens which initially invade the host at mucosal sites.

The *Lps* gene is one of four genes which have been shown to influence the susceptibility of mice to *Salmonella typhimurium* infection. The precise mechanism by which the *Lps* gene regulates immunity to *Salmonella*, however, is unknown. LPS-hyporesponsive C3H/HeJ (Lps^d/Lps^d) mice have been reported to be highly susceptible to *S. typhimurium* infection. The results of various studies have suggested that the failure of certain types of cells (including macrophages, T cells, and natural killer-like cells) to be activated or induced to differentiate by LPS may lead to defective biological activities of these cells, and may, therefore, account for the hypersusceptibility of C3H/HeJ mice to *Salmonella* infection. Thus, the C3H/HeJ mouse provides a useful model for further study of *Lps* gene regulation of susceptibility to *S. typhimurium* infection and for the investigation of the role of various antibody-mediated and cell-mediated immune mechanisms in resistance to *Salmonella*.

Acknowledgments. We wish to thank Drs. David E. Briles, William J. Koopman, John Volanakis (UAB), David L. Rosenstreich (Albert Einstein College of Medicine), and Stefanie N. Vogel (Uniformed Services University of the Health Sciences) for their helpful comments and criticisms, and Yvonne Noll for typing this review. The work from our laboratory which has been reviewed was supported by U.S. Public Health Service grants DE 02670, AI 19674, DE 04217, AI 18958 and DE 00092 and contract DE 42551. This review was submitted in partial fulfillment of the requirement of D.E.C. for the degree of Doctor of Philosophy in the Department of Microbiology in the Graduate School, The University of Alabama in Birmingham. Due to the continuous advancements being made in the area of *Lps* gene regulation of mucosal immune responses and resistance to *Salmonella* infection, it should be noted that this review summarizes work published prior to August 1984.

References

Andersson J, Coutinho A, Lernhardt W, Melchers F (1977) Clonal growth and maturation to immunoglobulin secretion in vitro of every growth-inducible B lymphocyte. Cell 10:27–34

Babb JL, McGhee JR (1980) Mice refractory to lipopolysaccharide manifest high immunoglobulin A responses to orally administered antigen. Infect Immun 29:322–328

Bockman DE, Cooper MD (1973) Pinocytosis by epithelium associated with lymphoid follicles in the bursa of Fabricius, appendix and Peyer's patches. An electron microscopic study. Am J Anat 136:455–478

Challacombe SJ, Tomasi TB Jr (1980) Systemic tolerance and secretory immunity after oral immunization. J Exp Med 152:1459–1472

Coutinho A, Meo T (1978) Genetic basis for unresponsiveness to lipopolysaccharide in C57BL/10Cr mice. Immunogenetics 7:17–24

Coutinho A, Möller G, Gronowicz E (1975) Genetical control of B-cell responses. IV. Inheritance of unresponsiveness to lipopolysaccharides. J Exp Med 142:253–258

Coutinho A, Forni L, Melchers F, Watanabe T (1977) Genetic defect in responsiveness to the B-cell mitogen lipopolysaccharide. Eur J Immunol 7:325–328

Dubos RJ, Schaedler RW (1960) The effect of the intestinal flora on the growth rate of mice, and on their susceptibility to experimental infections. J Exp Med 111:407–417

Eisenstein TK, Sultzer BM (1983) Immunity to *Salmonella* infection. Adv Exp Med Biol 162:261–296

Eisenstein TK, Deakins LW, Killar L, Saluk PH, Sultzer BM (1982) Dissociation of innate susceptibility to *Salmonella* infection and endotoxin responsiveness in C3HeB/FeJ mice and other strains in the C3H lineage. Infect Immun 36:696–703

Eisenstein TK, Killar LM, Stocker BAD, Sultzer BM (1984) Cellular immunity induced by avirulent *Salmonella* in LPS-defective C3H/HeJ mice. J Immunol 133:958–961

Glode ML, Rosenstreich DL (1976) Genetic control of B cell activation by bacterial lipopolysaccharide is mediated by multiple distinct genes or alleles. J Immunol 117:2061–2066

Gowen JW (1948) Inheritance of immunity in animals. Annu Rev Microbiol 2:215–254

Gowen JW (1960) Genetic effects in nonspecific resistance to infectious disease. Bacteriol Rev 24:192–200

Green DR, Gold J, St. Martin S, Gershon R, Gershon RK (1982) Microenvironmental immunoregulation: possible role of contrasuppressor cells in maintaining immune responses in gut-associated lymphoid tissues. Proc Natl Acad Sci USA 79:889–892

Hill AB, Hatswell JM, Topley WWC (1940) The inheritance of resistance, demonstrated by the development of a strain of mice resistant to experimental inoculation with a bacterial endotoxin. J Hyg 40:538–547

Hormaeche CE, Brock J, Pettifor R (1980) Natural resistance to mouse typhoid: possible role of the macrophage. In: Skamene E, Kongshavn PAL, Landy M (eds) Genetic control of natural resistance to infection and malignancy. Academic, New York, pp 121–130

Jensen SB, Mergenhagen SE, Fitzgerald RJ, Jordan HV (1963) Susceptibility of conventional and germfree mice to lethal effects of endotoxin. Proc Soc Exp Biol Med 113:710–714

Jirillo E, Kiyono H, Michalek SM, McGhee JR (1984a) Murine immune responses to *Salmonella* lipopolysaccharide: oral administration of whole bacteria to C3H/HeJ mice induces secondary anti-LPS responses, especially of the IgA isotype. J Immunol 132:1702–1711

Jirillo E, Michalek SM, McGhee JR (1984b) Lipopolysaccharide regulation of the immune response: LPS effects on lymphoreticular cells and mucosal immune responses. Estratto dalla rivista EOS IV:21–31

Johnson AG, Gaines S, Landy M (1956) Studies on the 0 antigen of *Salmonella typhosa*. V. Enhancement of antibody response to protein antigen by the purified lipopolysaccharide. J Exp Med 103:225–246

Joiner KA, McAdam KPWJ, Kasper DL (1982) Lipopolysaccharides from *Bacteroides fragilis* are mitogenic for spleen cells from endotoxin responder and nonresponder mice. Infect Immun 36:1139–1145

Kelly K, Watson J (1977) The inheritance of a defective lipopolysaccharide response locus in C3H/HeJ mice. Immunogenetics 5:75–84

Kiyono H, Babb JL, Michalek SM, McGhee JR (1980a) Cellular basis for elevated IgA responses in C3H/HeJ mice. J Immunol 125:732–737

Kiyono H, McGhee JR, Michalek SM (1980b) Lipopolysaccharide regulation of the immune response: comparison of responses to LPS in germfree, *Escherichia coli*-monoassociated and conventional mice. J Immunol 124:36–41

Kiyono H, McGhee JR, Wannemuehler MJ, Frangakis MV, Spalding DM, Michalek SM, Koopman WJ (1982a) In vitro immune responses to a T cell-dependent antigen by cultures of disassociated murine Peyer's patch. Proc Natl Acad Sci USA 79:596–600

Kiyono H, McGhee JR, Wannemuehler MJ, Michalek SM (1982b) Lack of oral tolerance in C3H/HeJ mice. J Exp Med 155:605–610

Kiyono H, Michalek SM, Mosteller LM, Torii M, Hamada S, McGhee JR (1982c) Enhancement of murine immune responses to orally administered haptenated *Streptococcus mutans*. Scand J Immunol 16:455–463

Lissner CR, Swanson RN, O'Brien AD (1983) Genetic control of the innate resistance of mice to *Salmonella typhimurium*: expression of the *Ity* gene in peritoneal and splenic macrophages isolated in vitro. J Immunol 131:3006–3013

McGhee JR, Mestecky J (eds) (1983) The secretory immune system. Ann NY Acad Sci 409:1–896

McGhee JR, Michalek SM, Moore RN, Mergenhagen SE, Rosenstreich DL (1979) Genetic control of in vivo sensitivity to lipopolysaccharide: evidence for codominant inheritance. J Immunol 122:2052–2058

McGhee JR, Michalek SM, Kiyono H, Babb JL, Clark MP, Mosteller LM (1982) Lipopolysaccharide regulation of the IgA immune response. In: Strober W, Hanson LA, Sell KW (eds) Recent advances in mucosal immunity. Raven, New York, pp 57–72

McGhee JR, Kiyono H, Alley CD (1984) Gut bacterial endotoxin: influence on gut-associated lymphoreticular tissue and host immune function. Surv Immunol Res 3:241–252

Michalek SM, Kiyono H, Babb JL, McGhee JR (1980) Inheritance of LPS nonresponsiveness and elevated splenic IgA immune responses in mice orally immunized with heterologous erythrocytes. J Immunol 125:2220–2224

Michalek SM, Kiyono H, Wannemuehler MJ, Mosteller LM, McGhee JR (1982) Lipopolysaccharide (LPS) regulation of the immune response: LPS influence on oral tolerance induction. J Immunol 128:1992–1998

Morrison DC, Ryan JL (1979) Bacterial endotoxins and host immune responses. Adv Immunol 28:293–450

Musson RA, Morrison DC, Ulevitch RJ (1978) Distribution of endotoxin (lipopolysaccharide) in the tissues of lipopolysaccharide-responsive and -unresponsive mice. Infect Immun 21:448–457

Nencioni L, Villa L, Boraschi D, Berti B, Tagliabue A (1983) Natural and antibody-dependent cell-mediated activity against *Salmonella typhimurium* by peripheral and intestinal lymphoid cells in mice. J Immunol 130:903–907

Nowakowski M, Edelson PJ, Bianco C (1980) Activation of C3H/HeJ macrophages by endotoxin. J Immunol 125:2189–2194

O'Brien AD, Rosenstreich DL (1983) Genetic control of the susceptibility of C3HeB/FeJ mice to *Salmonella typhimurium* is regulated by a locus distinct from known salmonella response genes. J Immunol 131:2613–2615

O'Brien AD, Scher I, Campbell GH, MacDermott RP, Formal SB (1979) Susceptibility of CBA/N mice to infection with *Salmonella typhimurium*: influence of the X-linked gene controlling B lymphocyte function. J Immunol 123:720–724

O'Brien AD, Rosenstreich DL, Metcalf ES, Scher I (1980a) Differential sensitivity of inbred mice to *Salmonella typhimurium*: a model for genetic regulation of innate resistance to bacterial infection. In: Skamene E, Kongshavn PAL, Landy M (eds) Genetic control of natural resistance to infection and malignancy. Academic, New York, pp 101–112

O'Brien AD, Rosenstreich DL, Scher I, Campbell GH, MacDermott RP, Formal SB (1980b) Genetic control of susceptibility to *Salmonella typhimurium* in mice: role of the *Lps* gene. J Immunol 124:20–24

O'Brien AD, Rosenstreich DL, Scher I (1981a) Genetic control of murine resistance to *Salmonella typhimurium* infection. In: Friedman H, Klein TW, Szentivanyi A (eds) Immunomodulation by bacteria and their products. Plenum, New York, pp 37–48

O'Brien AD, Scher I, Metcalf ES (1981b) Genetically conferred defect in anti-salmonella antibody formation renders CBA/N mice innately susceptible to *Salmonella typhimurium* infection. J Immunol 126:1368–1372

O'Brien AD, Metcalf ES, Rosenstreich DL (1982) Defect in macrophage effector function confers *Salmonella typhimurium* susceptibility on C3H/HeJ mice. Cell Immunol 67:325–333

Owen RL, Jones AL (1974) Epithelial cell specialization within human Peyer's patches: an ultrastructural study of intestinal lymphoid follicles. Gastroenterology 66:189–203

Plant J, Glynn AA (1976) Genetics of resistance to infection with *Salmonella typhimurium* in mice. J Infect Dis 133:72–78

Plant J, Glynn AA (1979) Locating salmonella resistance gene on mouse chromosome 1. Clin Exp Immunol 37:1–6

Robson HG, Vas SI (1972) Resistance of inbred mice to *Salmonella typhimurium*. J Infect Dis 126:378–386

Rosenstreich DL, Vogel SN (1981) The effects of endotoxin on macrophages and T-lymphocytes. In: Friedman H, Klein TW, Szentivanyi A (eds) Immunomodulation by bacteria and their products. Plenum, New York, pp 215–229

Schaedler RW, Dubos RJ (1961) The susceptibility of mice to bacterial endotoxin. J Exp Med 113:559–570

Schaedler RW, Dubos RJ (1962) The fecal flora of various strains of mice. Its bearing on their susceptibility to endotoxin. J Exp Med 115:1149–1160

Scher I (1981) B-lymphocyte development and heterogeneity. Analysis with the immune-defective CBA/N mouse strain. In: Gershwin ME, Merchant B (eds) Immunologic defects in laboratory animals, 1. Plenum, New York, pp 163–190

Scibienski RJ (1981) Defects in murine responsiveness to bacterial lipopolysaccharide. The C3H/HeJ and C57BL/10ScCr strains. In: Gershwin ME, Merchant B (eds) Immunologic defects in laboratory animals, 2. Plenum, New York, pp 241–258

Spalding DM, Koopman WJ, Eldridge JH, McGhee JR, Steinman RM (1983) Accessory cells in murine Peyer's patch. I. Identification and enrichment of a functional dendritic cell. J Exp Med 157:1646–1659

Sultzer BM (1968) Genetic control of leukocyte responses to endotoxin. Nature 219:1253–1254

Sultzer BM (1972) Genetic control of host responses to endotoxin. Infect Immun 5:107–113

Tagliabue A, Nencioni L, Villa L, Boraschi D (1984) Genetic control of in vitro natural cell-mediated activity against *Salmonella typhimurium* by intestinal and splenic lymphoid cells in mice. Clin Exp Immunol 56:531–536

Tomasi TB Jr (1980) Oral tolerance. Transplantation 29:353–356

Tomasi TB Jr (1983) Mechanisms of immune regulation at mucosal surfaces. Rev Infect Dis 5:S784–S792

Vogel SN, Rosenstreich DL (1979) Defective Fc receptor-mediated phagocytosis in C3H/HeJ macrophages. I. Correction by lymphokine-induced stimulation. J Immunol 123:1842–1850

Vogel SN, Hansen CT, Rosenstreich DL (1979) Characterization of a congenitally LPS-resistant athymic mouse strain. J Immunol 122:619–622

Vogel SN, Weinblatt AC, Rosenstreich DL (1981) Inherent macrophage defects in mice. In: Gershwin ME, Merchant B (eds) Immunologic defects in laboratory animals, 1. Plenum, New York, pp 327–357

von Jeney N, Günther E, Jann K (1977) Mitogenic stimulation of murine spleen cells: relation to susceptibility to *Salmonella* infection. Infect Immun 15:26–33

Wannemuehler MJ, Kiyono H, Babb JL, Michalek SM, McGhee JR (1982) Lipopolysaccharide (LPS) regulation of the immune response: LPS converts germfree mice to sensitivity to oral tolerance induction. J Immunol 129:959–965

Wannemuehler MJ, Michalek SM, Jirillo E, Williamson SI, Hirasawa M, McGhee JR (1984) LPS regulation of the immune response: *Bacteroides* endotoxin induces mitogenic, polyclonal, and antibody responses in classical LPS responsive but not C3H/HeJ mice. J Immunol 133:299–305

Watson J, Riblet R (1974) Genetic control of responses to bacterial lipopolysaccharides in mice. I. Evidence for a single gene that influences mitogenic and immunogenic responses to lipopolysaccharides. J Exp Med 140:1147–1161

Watson J, Riblet R, Taylor BA (1977) The response of recombinant inbred strains of mice to bacterial lipopolysaccharides. J Immunol 118:2088–2093

Watson J, Kelly K, Largen M (1978a) The genetic mapping of a locus in mice that controls immune and nonimmune responses to bacterial lipopolysaccharides. In: Friedman H, Linna TJ, Prier JE (eds) Infection, immunity and genetics. University Park Press, Baltimore, pp 25–38

Watson J, Kelly K, Largen M, Taylor BA (1978b) The genetic mapping of a defective LPS response gene in C3H/HeJ mice. J Immunol 120:422–424

Watson J, Largen M, McAdam KPWJ (1978c) Genetic control of endotoxic responses in mice. J Exp Med 147:39–49

Watson J, Kelly K, Whitlock C (1980) Genetic control of endotoxin sensitivity. In: Schlessinger D (ed) Microbiology – 1980. American Society for Microbiology, Washington DC, pp 4–10

Webster LT (1933) Inherited and acquired factors in resistance to infection. 1. Development of resistant and susceptible lines of mice through selective breeding. J Exp Med 57:793–817

Webster LT (1937) Inheritance of resistance of mice to enteric bacterial and neurotropic virus infections. J Exp Med 65:261–286

Williamson SI, Wannemuehler MJ, Jirillo E, Pritchard DG, Michalek SM, McGhee JR (1984) LPS regulation of the immune response: separate mechanisms for murine B cell activation by lipid A (direct) and polysaccharide (macrophage-dependent) derived from *Bacteroides* LPS. J Immunol 133:2294–2300

Genetic Determination of Bacterial Virulence, with Special Reference to *Salmonella*

B.A.D. STOCKER[1] and P.H. MÄKELÄ[2]

1 Introduction

"Virulence factor" or "virulence determinant" is often used to refer to bacterial traits, such as production of a surface component hindering phagocytosis, whose presence is noted to be correlated with virulence and whose loss, by mutation, etc., causes loss or great reduction in virulence in an experimental system. Logically any bacterial property indispensable for bacterial growth in the relevant compartment of the host, such as ability to grow at the temperature there prevalent, should be considered a virulence factor; we shall discuss such properties, as well as those commonly called virulence factors.

To survive in the tissue of a host, a pathogenic bacterium must be able to withstand the responses by which the host defends itself against microorganisms or other foreign material. One such defense is phagocytosis; it seems very likely that this evolved from the basic feeding behavior of eukaryotic organisms, as now seen in, for instance, the amoebae. In multicellular organisms, such as man and mouse, this defense is entrusted to specialized cells of several kinds,

[1] Department of Medical Microbiology, Stanford University School of Medicine, Stanford, California 94305, USA
[2] National Public Health Institute, Mannerheimintie 166, SF-00280 Helsinki, Finland

Current Topics in Microbiology and Immunology, Vol. 124
© Springer-Verlag Berlin·Heidelberg 1986

the professional phagocytes. To avoid being taken up by phagocytic cells, many kinds of bacteria have modified their basic or primitive cell surface in such a way that they are not recognized by the phagocytic cell, because of the particular properties of the protein, polysaccharide, or other macromolecular material which covers the easily recognized peptidoglycan cell wall substance. The hosts, however, have evolved by development both of the proteins of the complement system, able to recognize and bind to bacterial surface components not otherwise recognized, and of specialized complement receptors on the surface of the phagocytic cells, able to bind bacteria which have complement protein, especially C3b, deposited on their surface. Further development may then have been by the production by the bacteria of surface components which from their chemical nature do not bind complement (in the absence of specific antibody) and, by the host, evolution of the immune system, with the (inducible) production of immunoglobulin molecules able to bind to almost any bacterial surface component, including those which avoid direct activation of the complement system. Complement activation or binding could now be triggered by the antibody-antigen complex at the bacterial surface, so that bacteria could again be recognized by phagocytic cells, by means of the complement receptors on the surface of the phagocytic cell. In addition the evolution of Fc receptors would allow the direct recognition of bacteria with adsorbed immunoglobulins, without any requirement for binding of complement.

This scenario may be incorrect in some particulars, and we cannot be sure of the sequence of events in evolution. However, it illustrates the principle of continuous mutual evolution: there are always two agents, and their interaction, to be considered and successive strata of action/reaction underlying the present situation – the only one we can study directly. We may, by observation and experiment, discover the main factors now needed for a given bacterium to cause disease in a particular host; however, the course of evolutionary development of host defenses, and so of bacterial countermeasures, may well have differed in a different host species. Our observations and experiments on the *Salmonella* sp./mouse system have, however, revealed a variety of mechanisms involved in this system and many of them may be used by other bacteria as well. Such information allows some tentative generalizations, perhaps of value as a guide to action for prevention of infections; but we should be aware of their tentative nature and be prepared for surprises.

The best proof that a bacterial character is a virulence determinant is a comparison of the virulence (for a given host, etc.) of a pair of strains known of identical genotype except that only one has the character in question. If the character is well defined and easy to score, this comparison is often possible in pathogenic species in which methods of genetic analysis, for instance by transduction, are available; it may also be possible in other species it mutational gain, or loss, of the character can be observed. One may cite the proof of the role of the capsule of the Pneumococcus as a determinant of virulence for the mouse. The advantage of eliminating or minimizing host variability by use of inbred lines for experimental infections will be very evident from other chapters in this book. One serious limitation on experimental investigation of genetic aspects of bacterial virulence is the strict host-specificity of many important pathogens, e.g., of *S. typhi* for man (or higher primates). The physiological basis and genetic determination of host specificity remain completely unknown, and relatively little investigated, despite their evident importance in understanding the genetic basis of virulence. Because of absence of data we have nothing to discuss on this topic.

Bacteria of the genus *Salmonella* are important pathogens which have been extensively investigated, both in respect of genetics and as pathogens. Of the 2000-odd reported serotypes or species, only a very few have been much studied, because of their importance as causes of disease in man or farm animals. These include some which are "host-adapted," either infecting only a single host, for instance, *S. typhi* and man, or found mainly in association with one host – as *S. dublin* with bovines, and also others found in association with many different vertebrate hosts. *S. typhimurium*, the commonest species, has such a wide host-range and has been much investigated, for these reasons and because *S. typhimurium* infection in mice in many ways resembles typhoid fever in man; and we therefore take this species as the subject for consideration of factors identified as of importance for virulence.

2 Host as Habitat

2.1 Ability to Multiply in the Physical and Chemical Environment Provided by Host Tissue

Clearly a necessary condition for causing an infection in a given tissue of a particular host is the ability of the bacteria to survive and grow in the physical and chemical environment there encountered. Thus the appropriate temperature, oxidation-reduction potential, and concentration of inorganic components and of organic metabolites needed as nutrients for the pathogen concerned must be provided. (Their availability though essential is not sufficient for pathogenicity. Bacteria of many species which can multiply without restriction in a host environment, even from small inocula, fail to do so, or even to survive, unless they are protected from host cellular defenses, by enclosure in a cell-proof chamber; for instance, *Staphylococcus aureus* in the peritoneal cavity of the rabbit).

2.2 Temperature for Growth

We may consider first the effect of temperature. The failure of certain *Mycobacterium* sp. to produce other than superficial lesions in man, etc., very probably results from their inability to grow at a temperature as high as that found deeper within the body. In experimental *Treponema pallidum* infections in rabbits inoculation into the testicle or skin produces lesions, whereas injection into deeper tissues does not, probably because of the lower temperature in the more superficial situation. *S. typhimurium*, like many other common pathogenic species, grows well in vitro in suitable media at temperatures even as high as

44° C, which is higher than that encountered in mammalian tissues, even during high fever (whose evolutionary function may well be to retard or prevent multiplication of certain pathogens, bacterial or other, whose maximum temperature for growth is not much above normal body temperature). It might therefore seem that maximum temperature for growth was of little interest in relation to pathogenicity of *S. typhimurium*, etc. However, bacterial mutants called temperature-sensitive (for growth) though they grow well at, for instance, 30° C, fail to grow, or grow only for a generation or so, if they are transferred to temperatures as high as 37° C. In many cases at least the cause of this property is a mutation, probably a missense mutation in one codon, causing a single amino acid substitution, in the structural gene for a protein whose function is essential for growth, e.g., a DNA polymerase. Mutants of *Salmonella* sp. which are nonvirulent because they are temperature-sensitive for growth have been investigated for use as parenteral-route live vaccines, both in small laboratory animals and in calves (COOPER and FAHEY 1970; LINDE et al. 1974). The physiological basis of the temperature sensitivity of the mutants used and the genes affected by the mutations have been little investigated; however, LINDE et al. (1974) describe observations suggesting mutations affecting DNA polymerase as the probable cause in some strains of *Salmonella* tested as live vaccines. A nonvirulent strain to be used as a live vaccine should be known to be incapable of reversion to virulence, a requirement perhaps best met by showing that nonvirulence is due to deletion of part or the whole of a gene specifying a protein indispensable for virulence. Most temperature-sensitive mutations result from an alteration in the primary structure of a protein indispensable for growth such that the mutant protein functions at the lower temperature but not at the higher. The usual cause of this is a base-change mutation, and so reversible, either by true reversion (restoration of original codon and so of wild-type amino acid sequence) or by "second-site" mutation, in which the effect of the first amino acid substitution is partly or entirely reversed by a second amino acid substitution elsewhere in the same protein. Thus temperature-sensitive mutants are unlikely to be appropriate as live vaccines.

2.3 Oxidation-Reduction Potential and pH

Ability to multiply at the oxidation-reduction potential and partial pressure of oxygen prevailing in the host environment is presumably necessary for virulence – but little if anything is known about genetic determination of such ability. [The normal oxidation-reduction potential of the mammalian tissues apparently protect them against attack by obligate anaerobes such as *Clostridium perfringens*, which cannot initiate multiplication, and so infection, unless a tissue-damaging agent or vasoconstrictor (presumably causing reduced oxygen supply) is injected with the bacteria or at a site to which the bacteria can gain access.]

Ability to multiply at the pH prevailing in host tissues is also presumably essential for virulence – but here again there is little to be said about genetic determination of such ability. *Salmonella* sp. (and various other pathogenic bacteria) can multiply within host phagocytic cells, either because they can withstand the bactericidal conditions within the phagolysome or because they inhibit discharge of lysosomal material into the vesicle within which they lie. One such presumably bactericidal condition is the acid reaction produced by fusion with the lysosomes.

Exposure to gastric acidity appears to be an important defense mechanism against establishment of infection by ingested bacteria, as discussed below.

2.4 Iron Acquisition and Virulence

Injection of soluble iron compounds greatly promotes the ability to cause disease of inocula of many pathogenic or conditionally pathogenic bacterial species; this fact and the reversal by iron salts of the bacteriostatic effect of mammalian sera attest to the importance of the absence of assimilable iron as a factor controlling bacterial multiplication in host tissues, and suggest that bacterial mechanisms for iron acquisition may be essential for such multiplication, and so for virulence. Several mechanisms involving capture of iron by siderophores, either synthesized by the bacteria or from other sources, are known in the Enterobacteriaceae and in other groups. One such siderophore is enterochelin (= enterobactin), a cyclic trimeric ester of 2,3-dihydroxybenzoyl serine; it has an extraordinarily high affinity for ferric iron, so high that *Salmonella* sp. able to make enterochelin (and with the rest of the mechanism for binding of ferri-enterochelin and for transfer of its iron to the cell interior) can use iron bound to transferrin as the sole iron source (TIDMARSH and ROSENBERG 1981). Chromosomal genes, some clustered, involved in synthesis of enterochelin or with uptake and utilization of iron from ferri-enterochelin have been identified in *E. coli* and, to some extent, also in *Salmonella* sp. It might be expected that loss of any one of these gene functions by mutation would greatly reduce the virulence of a strain of *S. typhimurium*. YANCEY et al. (1979) isolated a mutant in the *enb* pathway (for synthesis of enterochelin from chorismic acid, the final product of the common aromatic biosynthesis pathway) in a moderately virulent *S. typhimurium* strain; this mutant was considerably reduced in virulence, as judged by i.p. DL_{50}. By contrast, BENJAMIN et al. (personal communication) found that the well-characterized *enb-1* and *enb-7* defects when transferred from *S. typhimurium* strain LT2 into a fully virulent *S. typhimurium* strain caused little if any increase in the LD_{50} for i.v. challenge. This suggests that nonavailability of iron may control multiplication of *S. typhimurium* in peritoneal fluid, and perhaps other intercellular fluids, but is not important as a factor controlling the multiplication of *Salmonella* within phagocytic cells in the liver and spleen (the relevant host compartment when bacteria are administered i.v.). A possible explanation is that the acid environment of the phagolysosome reduces the efficiency of host iron denial mechanisms* (see p. 172 for more recent information).

Bacteria of some species of Enterobacteriaceae use an alternative, or additional, iron uptake system; this comprises (at least) the ability to make and secrete a hydroxamate-type siderophore called aerobactin and to insert an aerobactin-binding protein in their outer membrane and to assimilate iron from ferri-aerobactin. WILLIAMS and WARNER (1980) showed that a conjugative plasmid, ColV-K30, determining production of colicin V and also increased (invasive) virulence in *E. coli*, included both genes for production of a siderophore (now known to be aerobactin) and others for transport of iron into the bacterial cell. The virulence-promoting effect of such plasmids in *E. coli* has been demonstrated; however, note that at least some such plasmids promote infection by another effect, viz., reducing susceptibility of their host to complement killing. The evolutionary advantage presumably conferred by possession of the aerobactin system as well as, or instead of, the enterochelin system is not known. However, some evidence suggests that aerobactin is more effective than enterochelin in reversing the bacteriostatic effect of normal mammalian sera, perhaps because of the presence of anti-enterochelin activity, thought to be antibody, in normal sera. *E. coli*, but not *S. typhimurium*, can use citrate as a sidererophore; the role, if any, of this system in infection is not known.

Any block in the common aromatic pathway prevents synthesis of enterochelin, an effect which in such mutants is reversed by provision of an intermediate of the pathway from chorismic acid to enterochelin, viz., 2,3-dihydroxybenzoic acid. The loss of virulence caused by blocks in aromatic biosynthesis, and the possible extent to which this loss results from interference with enterochelin synthesis, are discussed below.

In both *E. coli* and *S. typhimurium* efficient uptake of iron by any of the known siderophore systems depends on the function of a gene called *tonB* (in *E. coli*) or *chr* (*in S. typhimurium*). One might therefore suppose that loss of *chr* function would cause a considerable reduction in virulence, at least for the i.p. route. However, a proven deletion of all or part of gene *chr* transferred by two steps of transduction into a mouse-virulent strain of the same species caused, at most, a trivial reduction of virulence (S.K. Hoiseth and Stocker, unpublished observation).

2.5 Relation of Auxotrophic Character to Virulence

Bacteria of many pathogenic species, e.g., *Staphylococcus aureus*, in their wild-type form grow on defined media only if these are supplemented with one or more amino acid, purine or vitamin. Amino-acid-requiring mutants of initially nonexacting strains of *Salmonella* sp. correspondingly in general retain parental virulence; in nearly all cases reduced virulence of an amino-acid-requiring mutant when appropriately tested has been shown to result not from auxotrophic character but from coincident mutation or mutations (HERZBERG 1962). An exceptional case was encountered in an investigation of the effect of mutation to auxotrophy in a prototrophic (tryptophan-independent, cystine-independent) derivative of a strain of *S. typhi* (BACON et al. 1950a, b, 1951). Each of three induced auxotrophic mutants growing on defined medium only if it were supplemented with aspartic acid was of reduced virulence as tested in mice, i.p. injection, without adjuvant. (Note that in this model infection the dose required to produce a fatal infection is very large and death appears to result from extracellular multiplication). A recent reexamination of one of these mutants

(R. Brown and Stocker, unpublished) showed its mutation to be in gene *ppc*, specifying phosphoenolpyruvate carboxylase, an enzyme which bacteria use to refeed the tricarboxylic acid cycle with carbon skeletons to replace those withdrawn for synthesis of several different amino acids and other essential metabolites during growth on simple defined medium. A *ppc* derivative of a mouse-virulent *S. typhimurium* strain was found to be about as virulent as its parent strain, perhaps because *ppc* derivatives of *S. typhimurium* (unlike the *S. typhi* mutants) could utilize citrate, or various other tricarboxylic acid cycle compounds, presumably available in sufficient concentration in host intracellular fluids. Thus, ppc^+ could be considered a virulence determinant for *S. typhi* (at least in the animal model used) though not in *S. typhimurium*. As discussed below a block in the common aromatic biosynthesis pathway causes requirement for, among other aromatic metabolites, the three aromatic amino acids; such biosynthetic blocks cause nonvirulence in *Salmonella* sp., but apparently because of the requirement for aromatic "vitamins" rather than as a consequence of amino acid auxotrophy.

As long ago as 1950 it was observed that mutation causing requirement for a purine, or for both a purine and vitamin B1 (the expected consequence of an early block in the purine biosynthesis pathway), caused an increase in the LD_{50} of *S. typhi*, tested in mice, i.p. route, without adjuvant (BACON et al. 1950a, b, 1951). Reduced virulence of purine auxotrophs has since been noted in other species and other genera and, in some instances, proven to be a result of the auxotrophy by demonstration of increased virulence of purine-independent derivatives, revertant or transductant, or by the infection-promoting effect of parenteral administration of an appropriate purine, etc. The mechanism of reduction of virulence is presumed, in some instances proven, to be the insufficient concentration of an available source (base, nucleoside, or perhaps nucleotide) of the relevant purine in the host compartment concerned, i.e., in intercellular fluids or, in the case of intracellular parasites, within phagocytic cells. This inadequate supply presumably reflects the fact that the free bases and nucleosides are not intermediates in the de novo synthesis of purine nucleotides, which can be effected by most mammalian cells. However, mammalian cells can also salvage purines and nucleosides derived from breakdown of cellular nucleic acid or from dietary sources and so presumably will contain at least some of the substrates of these pathways; furthermore, cells of some host tissues, for instance, erythrocytes, cannot synthesize purines de novo and must rely entirely on salvage of purines reaching them via plasma. Thus reduced virulence of purine-auxotroph mutants of bacteria of pathogenic species normally able to make their own purines may reflect inability of these bacteria to take up the relevant purine at a rate sufficient to allow multiplication when the ambient concentrations is very low. It is noteworthy that all investigated parasitic protozoa, including those which multiply in the plasma component of blood, are purine auxotrophs (GUTTERIDGE and COOMBES 1977), as also are strains of *Neisseria gonorrhoeae* of the type apt to cause generalized infections; presumably these organisms have evolved very efficient purine uptake systems.

In *E. coli* and *S. typhimurium*, the best-investigated species, a block in any of the eight steps of the de novo pathway before the branch-point compound,

IMP (inosinemonophosphate, the nucleotide form of hypoxanthine), results in a purine requirement sastisfied by provision of adenine, guanine, xanthine, or hypoxanthine or of an available compound of any of them; a block at *purA* and *purB*, i.e., in either of the two steps for conversion of IMP to AMP, causes requirement specifically for adenine (or adenosine, etc). and, correspondingly, blocks between IMP and GMP cause specific requirement for guanine or the relevant intermediate, xanthine. These different requirements may have different effects on virulence. Thus IVANOVICS et al. (1968) found that *Bacillus anthracis* mutants with blocks between IMP and AMP were nonvirulent for mice, whereas those with blocks before IMP retained virulence. BRUBAKER (1972) found that in *Yersinia pestis* mutations before IMP had little if any effect on mouse virulence but that mutations affecting conversion of IMP to GMP, and so causing requirement for guanine, resulted in complete loss of (mouse) virulence. In a recent investigation (W. McFarland and Stocker, unpublished) of the effect of purine auxotrophy on virulence of a Vi-positive *S. dublin* strain for an Ity-s mouse line, BALB/c, (i.p. route), transductants with transposon Tn10 inserted in genes for early steps in the pathway in some instances retained considerable virulence, as shown by the (late) deaths of some mice given inocula of $<10^3$ colony-forming units (CFUs). The same was observed for derivatives with blocks between IMP and GMP; by contrast complete blocks between IMP and AMP resulted in nonvirulence (no illeffects from i.p. inocula of 10^6 CFUs, compared with an LD_{50} of <10 for the nonexacting parent strain). Presumably these results reflect differences in the concentration of various available purine source in host intercellular fluids and/or within phagocytic cells.

The effect of mutations causing requirement for any pyrimidine, or for a specific pyrimidine, have been less fully investigated. However, SMITH and TUCKER (1976) recorded a considerable reduction, but not complete loss of virulence, in thymine-requiring mutants of *S. typhimurium* – a class which are easy to isolate because of their resistance to trimethoprim; the same observation has been made for other strains of *S. typhimurium* made *thyA* by transduction (C. Spurdon and Stocker, unpublished).

A mutant of *Salmonella typhi* with a requirement for *p*-aminobenzoate (pAB) was of reduced virulence, as indicated by an increased LD_{50} in the mouse, i.p. route, without adjuvant, system (BACON et al. 1950a, b, 1951); administration of pAB in depot form reversed the reduction in   virulence. In *E. coli* and *Salmonella* sp., pAB can be taken up from the exterior, or made (by a reaction catalyzed by the products of genes *pabA* and *pabB*) from chorismic acid, the final product of the aromatic (*aro*) biosynthesis pathway. In these and many other bacterial species, and in plants, pAB is the precursor of folic acid. Vertebrates cannot synthesize folic acid, which they instead acquire ready-made, as a dietary component or vitamin; *E. coli* and *Salmonella*, by contrast, cannot assimilate folic acid and must make it themselves, by joining pAB, a pteridine and glutamic acid. pAB has no metabolic role in vertebrates; adsorbed or injected pAB is rapidly conjugated and excreted. It is therefore expected to be absent, or present in only trace amounts, in vertebrate tissues (a situation which explains the efficacy of sulfonamide treatment of many bacterial infections

– if pAB were present in plasma in the same concentration range as are sulfona-
mides during chemotherapy, the bacteriostatic effect of the latter would be
reversed).

A complete block in synthesis of chorismate by transposon insertion (and
secondary mutation) at gene *aroA* caused virtually complete loss of ability to
cause systemic infection in mouse-virulent strains of *S. typhimurium* and *S.
dublin* (HOISETH and STOCKER 1981; STOCKER et al. 1983; SMITH et al. 1983,
1984; ROBERTSSON et al. 1983). The *aro* genes needed for aromatic biosynthesis
are thus virulence genes or determinants in this genus. It is not known whether
the nonvirulence of aromatic-dependent strains results only from their requirement for pAB or in part also from their requirement for 2,3-dihydroxy-benzoic
acid (for synthesis of enterochelin) or for other minor products of chorismate,
in particular the precursors of ubiquinone and the menoquinones. pAB though
absent, or nearly so, from vertebrate tissues is present in many other environments, including the gut lumen – since *S. typhimurium* which are nonvirulent
because of *aro* mutation reach high levels in the gut of germ-free chicks (G.H.
Snoeyenbos, personal communication) or mice (receiving streptomycin by
mouth, to remove most of the normal flora, which would otherwise impede
colonization by *Salmonella* sp.) (NEVOLA et al. 1985). Thus, ability to grow
without exogenous pAB though a virulence factor in respect of invasive infection
would probably not be so far a bacterial species causing disease by toxin production within the gut lumen, e.g., enterotoxin-producing *E. coli*.

3 Bacterial Reaction to First-Line (Chemical) Host Defenses

In most animals general (chemical) defenses against bacterial invasion include
the adverse pH and surface-active bile salts met in the digestive tract. As noted
above the acidity of gastric juice kills most ingested bacteria. Evidently some
enteric pathogens survive gastric passage, perhaps because of the occasional
very rapid transit of ingesta into the duodenum or because of shielding or
buffering by food; or perhaps infection occurs mainly in hosts with reduced
acid production. Special anaerobic bacteria residing on the human gastric mucosa have recently been described (LANGENBERG et al. 1984), but their mechanism of resistance to acid is not known. No specific defenses against exposure
to acid are known in the *Salmonella* group, which, in the case of *S. typhimurium*,
must also survive (and multiply in) the acid environment of the phagolysosome
if it is to cause mouse typhoid (O'BRIEN et al. 1979). It is of interest that an
obligate intracellular pathogenic bacterium, *Coxiella burnetii*, the only species
of this genus to multiply within mouse phagolysosomes rather than in the cytoplasm, is reported to show metabolic activities when tested at pH 2.0–5.5, in
contrast to inactivity at neutral pH (HACKSTADT and WILLIAMS 1981).

Bile acids or salts are detergents and solubilize many biological membranes,
yet many gram-negative bacteria not only survive but multiply in the gut. It
is now clear that detergent resistance depends on an appropriate structure of

the bacterial outer membrane. Many nonenterobacterial gram-negative bacteria, including *Neisseria* and *Hemophilus* sp. (LYSKO and MORSE 1981), though they have an outer membrane are killed by bile salts – as are bacteria of some normally resistant species if they have mutations altering outer membrane structure in such a way as to cause detergent sensitivity. In particular, *Salmonella* mutants making lipopolysaccharide (LPS) with a deep core defect are sensitive to both anionic and cationic detergents and to various other cell-damaging agents (VAARA and NIKAIDO 1984; SCHLECHT and SCHMIDT 1969) including some dyes, hydrophobic antibiotics, and the cationic polypeptide of phagocytes (MODRZAKOWSKI and SPITZNAGEL 1979; WEISS et al. 1983), as discussed below. This may result from the presence of phospholipids in the outer leaflet of their outer membranes (VAARA and NIKAIDO 1984). A common feature of these deficient LPSs is absence of the phosphates normally linked to the heptose residues in the inner part of the core (DROGE et al. 1968); this suggests that ionic interactions between outer membrane components, either between LPS molecules or of LPS with protein, are important for detergent resistance. Some mutations affecting non-LPS outer membrane components (for the most part unidentified) also affect detergent resistance, whose genetic determination is thus multifactorial (SCHWEIZER et al. 1976; COLEMAN and LEIVE 1979; NORMARK 1969; SUKUPOLVI et al. 1984).

4 Response to the First-Line Mechanical Defenses; Adhesion and Invasion

The peristaltic movement of the intestine tends to pass on and remove its contents, including potentially invading bacteria. A similar cleansing function is effected by the flow of urine in the urinary tract, and by movement of mucus by ciliary action on ciliated mucosal surfaces. In this situation the bacteria need to hold fast, and attach themselves to the mucosal surface. There is good evidence for the role of specific attachment organelles, called fimbriae, in the binding of the bacteria of various pathogenic species to specific receptors on the surface of cells in the intestine or the urinary tract. Thus fimbriae called K88 or K99 (on the basis of their reactions with antisera) of *E. coli* are determined by plasmid genes (in fact by a cluster of at least six cistrons) (GAASTRA and DE GRAAF 1982). While bacteria possessing such fimbriae adhere to cells, colonize the intestine, and cause diarrhea in piglets, their isogenic variants, plasmidless or with a mutation in one of the fimbrial genes, therefore lacking fimbriae, do not adhere or cause diarrhea (SMITH and LINGGOOD 1971; JONES and RUTTER 1972; MOON et al. 1977). In pigs the presence or absence of the receptor for K88 fimbriae is under genetic control and pigs whose intestinal mucosal surfaces lack the receptor are not susceptible to diarrhea caused by K88 *E. coli* (SELLWOOD et al. 1975). Furthermore, antibody to the fimbriae in the colostrum taken by piglets prevents such diarrhea (RUTTER and JONES 1973). In *E. coli* the presence of specific fimbriae – called P because of their binding to P-blood-group-specific receptors on human red cells (KÄLLENIUS et al. 1980; LEFFLER and SVANBORG-EDEN 1980) – with the ability of *E. coli* strains to cause pyelonephritis in man is associated with ability to cause pyelonephritis in man (VÄISÄNEN et al. 1981). Gene clusters determining synthesis of P fimbriae have been cloned (HULL et al. 1981; CLEGG 1982; NORMARK et al. 1983; RHEN et al. 1983a). Although chromosomally located, the gene clusters concerned turn out to be very similar to the plasmid-borne K88 and K99 gene clusters, both in their internal organization and in the polypeptides which they code (NORMARK et al. 1985).

Enteric bacteria have many other types of fimbriae with different, only partially elucidated, receptor specificities. In fact, a single *E. coli* strain can express at least four different kinds of fimbriae (NOWICKI et al. 1984). However, their putative role in pathogenesis is poorly understood. α2-linked sialosyl residues are common on mammalian cells; S-fimbriae binding to sialyl α2–3 galactoside have been demonstrated more often in *E. coli* strains isolated from septic infections of newborn infants than in strains from other sources (PARKKINEN et al. 1983; KORHONEN et al. 1985). Mannose is likewise a common constituent of mammalian cell-surface glycolipids and glycoproteins, and most enteric bacteria can make so-called common or type-1 fimbriae that bind to α1-linked mannosyl residues (DUGUID and GILLIES 1958; OLD 1972). The possible association of such fimbriae with pathogenicity has, however, been elusive. It has been suggested that in urinary tract infections they may bind the bacteria to the mucus overlying the bladder mucosa and thus, in fact, prevent access of the bacteria to the cells (ØRSKOV et al. 1980). Type-1 fimbriae are the only kind of fimbriae demonstrated in *Salmonella*; a weak association with virulence was shown in a study in which the fimbriate *S. typhimurium* was slightly better than its nonfimbriate sister strain at initiating infection after feeding to mice (DUGUID et al. 1976).

A feature of fimbriae is their variable expression. Thus a given temperature of incubation or a particular nutrient medium may totally suppress their production (SAIER et al. 1978; DE GRAAF et al. 1980; GIRARDEAU et al. 1982; GÖRANSSON and UHLIN 1984). More unusual, however, is the rapid on-off switching of their production, shown both for type-1 fimbriae and for several fimbrial types of *E. coli* from urinary tract infections; this occurs in a coordinated fashion, so that most cells in a culture at any one time have only one type of fimbriae (BRINTON 1959; RHEN et al. 1983b; NOWICKI et al. 1984). The genetic mechanism of these switches is now being studied; the fimbrial gene clusters have been cloned and, in large part, sequenced but on/off variation is difficult to demonstrate when the genes are carried on multicopy plasmids (EISENSTEIN 1981; RHEN et al. 1983a; ORNDORFF and FALKOW 1984). One may imagine that variation is useful to the bacteria during the course of an infection, for instance, by allowing the bacteria to change their affinity for different host-cell surfaces, according to their location during the course of an infection. However, so far such possibilities have not been studied experimentally.

Other, nonfimbrial, adhesins mediating attachment of bacteria to host cells are also present in various species of enteric bacteria. In *Salmonella typhimurium* an adhesin mediating agglutination of glutaraldehyde-fixed sheep red cells has been shown to be determined by a cluster of chromosomal genes which have been cloned in *E. coli* and shown to confer ability to adhere to mammalian cells (MADELON HALULA and STOCKER 1984, 1985). The possible role of this mechanism in pathogenesis remains to be determined. In *E. coli* one such non-fimbrial adhesin was shown to be of molecular weight ca. 16000, as are the type-1 and P fimbrial adhesins. However, neither the nonfimbrial protein nor the gene specifying it could be shown to be closely related to its fimbrial analogues (LABIGNE-ROUSSEL et al. 1984). In fact there is genetic evidence that the P-specific attachment associated with presence of P fimbriae is mediated not by the fimbrial protein itself but by a polypeptide coded by another gene in the same cluster (NORMARK et al. 1983) – in this case the fimbriae may serve as organelles which present the adhesin effectively.

In the *Yersinia* group a virulence-associated plasmid codes for a protein which forms projections much shorter than typical fimbriae (described as tack-like) on the bacterial surface (ZALESKA et al. 1985). This protein mediates adhesion of the bacteria to each other and to host cells (SKURNIK et al. 1984). Its expression is controlled by temperature of growth, like that of many other fimbriae (BÖLIN et al. 1982).

Invasion, i.e., penetration, into host cells or through the epithelium is the next step after binding in the pathogenesis of many infections. The invasive ability of *Shigella* and *Yersinia* species has been shown to depend on the presence of a large plasmid (ZINC et al. 1980; GEMSKI et al. 1980; PORTNOY et al. 1983; SANSONETTI et al. 1982, 1983). The products of the plasmid genes involved and the actual mechanism of invasion are still unclear, however. The *Shigella* multiply in the epithelial cells without invading through the mucosa, whereas *Salmonella* get into epithelial cells of the small intestine and proceed through them to the submucosa, as shown by electron microscopy (TAKEUCHI 1967). The requirements for these events and their regulation are not yet known in detail.

5 Coping with the Complement System

The complement cascade is activated by many foreign structures, including polysaccharides. This activation is commonly via the "alternative pathway," by a direct binding of the C3 component, leading to its cleavage to C3b (which becomes covalently linked via a thioester linkage to a suitable nearby structure) and expansion of the cleavage events (BJÖRNSON and BJÖRNSON 1977; LAW et al. 1979; PANGBURN and MÜLLER-EBERHARD 1980). Activation may also occur by direct binding of C1q and initiation of the classical pathway; for instance, the lipid A part of LPS does this (MORRISON and KLINE 1977) and so do some types of smooth LPS (PLUSCHKE and ACHTMAN 1984; TENNER et al. 1984). An obvious way for the bacteria to avoid the deleterious effects of activated complement would be avoidance of the activation step. Many kinds of bacteria have found ways to accomplish this. The means used seems to be covering of the otherwise activating surface (lipid A, LPS core, etc.) by a thick layer of nonactivating polysaccharide, either a suitable type of O polysaccharide (covalently linked to the LPS core) or capsular polysaccharide. A comparison of isogenic derivatives of *Salmonella typhimurium* differing in the structure of their O polysaccharide showed that they differed greatly in their ability to activate complement via the alternative pathway, and that this difference was correlated with their differing ability to resist phagocytosis and differing virulence for the mouse (LIANG-TAKASAKI et al. 1982, 1983; VALTONEN 1970, 1977). Derivatives of the same strains without any O polysaccharide were nonvirulent and easily phagocytosed, and activated complement rapidly (VALTONEN 1970; GROSSMAN and LEIVE 1984). Different capsular polysaccharides of *E. coli* differ in their ability to activate complement, with poor ability to activate corresponding approximately to frequency in invasive infections (STEVENS et al. 1983), but the relevant isogenic derivatives were not tested. Some capsular polysaccharides are completely inactive toward complement, thus even less active in this respect than the above-mentioned O polysaccharides of *Salmonella*. These include the already-mentioned polysialosyl K1 polysaccharide, whose structure resembles that of certain oligosaccharides in human glycoproteins (FINNE et al. 1983). When isogenic derivatives of an encapsulated *E. coli* O18:K1 strain were compared and those devoid of either O polysaccharide or of capsular K1 polysaccharide were tested for virulence in chick and mouse, the presence or absence of the capsule was found decisive for virulence (SMITH and HUGGINS 1980). However, complement activation was not tested. In a different series of experiments, the capsular polysaccharides (of type K1) were shown to prevent the alternative-pathway activation that occurred with the noncapsulate $18:K^-$ mutants (PLUSCHKE et al. 1983a). Another mutant which had lost the O polysaccharide ($O^-:K1$) activated complement via the classical pathway, without the need for antibodies. The same was true of strains with one type of O polysaccharide (O1:K1) but not true for strains with another combination (O7:K1). The complement activation paralleled ability to cause bacteremic infection after feeding to newborn rats (PLUSCHKE et al. 1983b).

Still another way for bacteria to cope with the complement system is to resist the effects of activated complement. This approach is used by many bacteria of the enteric group. The end product of complement activation is the membrane-attack complex (MAC), formed from several of the last components of the complement cascade. The ring-shaped MAC inserts into membranes (artificial, eukaryotic, or procaryotic) and causes lysis of the membrane, vesicle or cell, because of its central hole or channel (BHAKDI et al. 1980; HAMMER et al. 1977; SCHREIBER et al. 1979; WRIGHT and LEVINE 1981). The bactericidal action of complement takes place in this way, by the insertion of the MAC in the outer membrane (gram-positive bacteria, in which the only cell membrane is under the thick peptidoglycan layer, are resistant to this action of complement). How can enteric bacteria resist this action of the MAC?

A very clear-cut answer to this question was given recently by the elegant studies of JOINER et al. (1982a, b). In brief, this resistance involves letting the process of complement activation take place in a hydrophilic environment where the MAC cannot insert. The O polysaccharide of smooth *Salmonella* provides such an environment, where MAC does not have direct enough access to the surface of the outer membrane. By contrast, in a mutated derivative of the O polysaccharide, the insertion of MAC was efficient, and the cells were killed.

Not all O polysaccharides give equally good protection against the insertion of the MAC; some smooth *E. coli* strains are sensitive to killing by complement, but can be made resistant by a mutation altering the distribution (perhaps density) of the O polysaccharide (JOINER et al. 1984).

The integrity and composition of the outer membrane also influence sensitivity to killing by complement. Thus the outer membrane protein, TraT, determined by a gene of conjugative plasmids increases the resistance of their bacterial host (MOLL et al. 1980) whereas binding of polycationic peptides decreases it (VAARA and VAARA 1983; VAARA et al. 1984). Another effect of complement activation is "opsonization," i.e., sensitization of the bacteria to phagocytosis. In the next section we discuss how some bacteria deal with this effect.

6 Coping with Phagocytosis

Phagocytosis must start by sufficiently close contact between the phagocytic cell and the particle, in this case the bacterium, to be attacked. For this purpose the phagocytes have several sorts of receptor on their surface; those for immunoglobulin (Fc receptors) and for complement components, including the C3b receptor, are the most important in this context, and also the best studied. Once again it is important to the bacteria that they avoid activating complement, because the deposition of C3b generated in the activation process would tag them for binding to the C3b receptor. Thus, O polysaccharide and the capsular polysaccharide form the basic shield to protect the bacteria against phagocytosis, and their usefulness for this purpose depends on their ability to avoid activating complement. The K1-type polysialosyl capsule seems to be ideal; it is often referred to as antiphagocytic, with the implication that it not only fails to cause deposition of C3b but that it actively repels the phagocytic cells. What this means in molecular terms is, however, not specified, and this mechanism is not proven.

Binding to the receptors of the phagocytes is followed by ingestion, which, however, does not always occur. Thus binding to the C3b receptor does not give the phagocyte the required signal for ingestion unless the Fc receptor also is occupied (EDWANG and BEFUS 1984) or the phagocyte has been prepared, for instance, stimulated by thioglycolate (GRIFFIN and MULLINAX 1981). How the bacteria exploit these requirements for ingestion (or whether they do so) has not so far as we know been specifically investigated.

The binding of the bacteria to the phagocyte also signals it to activate its killing mechanisms. One such is the oxygen-radical pathway – a respiratory burst creates a series of active derivatives of oxygen in the phagocytic vesicles and also outside the cell (BABIOR 1978). This may be a means for killing the bacteria without ingesting them, and there is evidence for such extracellular killing (BLUMENSTOCK and JANN 1981). The enteric bacteria, like most other cells, seem to be quite sensitive to this killing mechanism, and even smooth *Salmonella*, which are resistant to most other host defenses, are killed by it.

Another means phagocytes have for killing intracellular bacteria is with the mixture of lysosomal enzymes and cationic proteins which they discharge into the phagosome (ELSBACH and WEISS 1983). The environment in this compartment is acidic, pH 4.5–5.0, which in itself is harmful to the enteric bacteria. Most enteric bacteria are unable to survive this attack, but smooth *Salmonella* are resistant. Studies with isolated lysosomal proteins have shown that the main *Salmonella*-cidal action is caused by the cationic peptides; the O polysaccharide is important for the resistance of smooth bacteria, and mutants with defective LPS core are very sensitive (MODRZAKOWSKI and SPITZNAGEL 1979; WEISS et al. 1983).

The main multiplication of *Salmonella* in mouse tissues, in fact, takes place in the macrophages of the liver and spleen (COLLINS 1969); the more virulent the strain, the more rapid is such bacterial multiplication. We are thus faced with the paradox that the *Salmonella* seem to have evolved elaborate O polysaccharide structures to avoid activating complement, yet finally depend on getting into phagocytic cells and multiplying in them. The explanation appears to lie in the kind or quality of the phagocytic cells concerned – those in the peritoneal cavity rapidly kill *Salmonella* tagged with C3b, whereas those in the spleen and liver do not (SAXÉN 1984; SAXÉN et al. 1984). Thus the bacteria need to avoid complement activation in the vicinity of killer phagocytic cells but complement activation is acceptable, perhaps even beneficial, in other locations. The characteristics of the different sorts of phagocytes remain to be described before this phase of *Salmonella* infection is adequately understood.

Smooth *Salmonella* (with a complete O polysaccharide) and also their rough mutants having complete core LPS but no O polysaccharide survive and multiply in liver and spleen macrophages, whereas mutants with incomplete core are rapidly killed (SAXÉN et al. 1984). All the *Salmonella* genes affecting LPS structure, and thus likely to affect ability to grow in macrophages, are located in the chromosome; however, the large "cryptic" plasmids found in some but not all *Salmonella* serotypes also affect intracellular survival and growth. In *S. typhimurium* in which the role of the cryptic plasmid in virulence was discovered, it seems that the step concerned is adhesion and/or invasion (JONES et al. 1982). However, in some recent experiments involving this species (M.F. Edwards and Mäkelä, unpublished) the presence or absence of the plasmid appeared to affect virulence by affecting the behavior of the bacteria within phagocytic cells.

Yersinia sp. can also survive and grow in macrophages, and this property likewise is dependent on their large "virulence plasmids" (BERCHE and CARTER 1982; SIMONET et al. 1984). The Vw protein has been suggested as an essential requirement for the intracellular growth of *Y. pseudotuberculosis* ssp. *pestis*, but the mechanism of action is still unclear (PORTNOY et al. 1983).

7 Immune Defense: Do the Bacteria Still Have a Chance?

Antibodies to the O polysaccharide or the capsular polysaccharide can obviously annul the capacity of these structures to avoid activating complement. A possible escape for the bacteria would be a capacity to alter their surface structure, during or after an infection, so that antibodies evoked by the original form would not recognize the new surface. Some bacteria (for instance, the relapsing fever spirochete) and some eukaryotic parasites do just this, but there is no definite evidence that enteric bacteria do so. However, these bacteria do have mechanisms for changing their surface structures – the flagellar antigen undergoes phase variation (production of flagella made from one flagellin protein or from an alternative protein) and smaller alterations in the structure of capsular polysaccharide and O polysaccharide, called form variation, are known (MÄKELÄ and STOCKER 1984; LEDERBERG and IINO 1956; ZIEG et al. 1977; KAUFFMANN 1941; MÄKELÄ and MÄKELÄ 1966; ØRSKOV et al. 1979). However, we do not know what benefits the bacteria derive from these variations, though possible benefits have been suggested for variation of fimbrial type. The changes are not very extensive, and their rate rather low, which makes their role in avoiding host antibody response less likely.

The antibodies bound to the bacteria not only activate complement but also provide binding sites for the Fc receptors of phagocytes. Thus phagocytosis, including rapid ingestion, can be expected to be efficiently promoted, and bacteria sensitive to phagocytic killing to be destroyed. By contrast, Salmonella are only marginally affected, in respect of intracellular fate, by antibodies, in keeping with their capacity for intracellular growth. Passively administered antibodies indeed have no effect on intracellular growth, and thus on overall multiplication, of S. typhimurium during the greater part of an experimental infection (COLLINS 1969; SAXEN 1984). The antibodies did have a small effect in a short first phase of infection after intravenous injection of the bacteria. This effect was much greater (but still confined to the first hour or so) after intraperitoneal injection – and remember that peritoneal phagocytes can kill complement-coated Salmonella, whereas liver and spleen macrophages cannot (BLANDEN et al. 1966; SAXEN 1984).

However, T-cell-mediated immune mechanisms can activate macrophages to kill even intracellular Salmonella, and so at this stage the host wins (MACKANESS et al. 1966). The development of effective immunity takes quite some time – unimpaired growth of Salmonella in the liver of the mouse can be seen for 7–10 days before the bacterial numbers start declining (COLLINS 1969; MACKANESS et al. 1966). Thus bacteria that can grow to a sufficient population size (which in mouse salmonellosis is approximately 10^8 organisms in the liver) in this time period will win, and kill the animal, and this is exactly what virulent Salmonella do. On the other hand, we have a chance, too: we can stimulate T-cell immunity beforehand and thus reduce to the minimum the time during which the bacteria can multiply. This can best be done by administration of live vaccine (HOISETH and STOCKER 1981). In hosts immunized in this way, the number of bacteria surviving in the liver and spleen remain stationary or begin to decline within a day or two after intravenous challenge (data to be published).

8 Genetics of Bacterial Toxin Production

The pathogenic ability of many bacterial species is dependent on their production of toxins. For bacteria for which infection is a "dead end," and therefore without significance in bacterial evolution, the toxicity of some bacterial product for animal cells or tissues is presumably an accidental consequence of some activity, enzymic or other, of a bacterial product with some (usually unknown) function needed for multiplication of the bacteria in their normal habitat. Such is presumably the case for the *Pseudomonas aeruginosa* exotoxin, which has the same mode of action as diphtheria toxin, i.e., enzymic addition of an ADP-ribosyl group to a specific amino acid in a specific protein involved in protein synthesis. In *Corynebacterium diphtheriae*, by contrast, the production of toxin is clearly needed for production of diphtheria; the infrequent isolation of even nontoxigenic *C. diphtheriae* from well-immunized populations suggests that strains of the latter sort cannot maintain themselves in a human population (the ecological niche of this species) and, one may infer, that evolutionary pressure selects for continued toxigenicity, and so virulence, in this species. One can only speculate as to how the structural gene for diphtheria toxin came to be located in a prophage genome.

The toxic (endotoxin) effects of LPS, dependent as they are on its lipid A component, are characteristic of the LPS of gram-negative bacteria of many free-living species, and are presumably a consequence of a structure evolved to meet the requirements of existence in such habitats and in a sense accidental, in respect of role in pathogenesis.

Of the relatively few exotoxins made by Enterobacteriaceae the best characterized are perhaps the stable (ST) and the heat-labile (LT) enterotoxins, made by some *E. coli* strains. The LT toxins of *E. coli* resemble the toxin made by *V. cholerae* in toxin structure (two polypeptides, one for adsorption of the compound molecule to a receptor in gut epithelial surface, the other an enzymically active component which enters the cell) but differ in that the *E. coli* enterotoxin genes are found in plasmids, those of *V. cholerae* in the chromosome. In *V. cholerae* serial passage in gut loops yields variants which make much more toxin than their parent strain (MEKALANOS 1983); these arise by amplification, i.e., the production of many tandem repeats of a chromosome segment which includes the toxin genes. Analogy with amplification in other systems suggests that this occurs by repeated unequal crossing over between copies, in sister chromosomes, of a repeated sequence, originally present as a tandem duplication. Such duplications are now known to arise continuously, and to be eliminated, at fairly high frequency in *S. typhimurium* (ANDERSON and ROTH 1977).

The toxin produced by *Shigella shigae* type 1 – and, it now appears, also by many other enterobacterial species, including some strains of *E. coli* – is of particular interest both because of the possibility that its production by some *E. coli* strains is related to their ability to produce diarrheal disease and because of the discovery of phages which can "convert" *E. coli* to high production of Shiga-like toxin (O'BRIEN et al. 1984); it is not yet known if the phages carry the structural gene for the toxin or whether they somehow cause strong expression of toxin structural genes already present in the chromosome of the recipient but previously not expressed or expressed only weakly.

Hemolysin production may contribute to virulence in *E. coli*, as indicated both by frequency of hemolysin production among isolates from certain infections and from the results of model infections in animals. Clusters of genes needed for hemolysin production occur either in the chromosome or in plasmids in *E. coli*.

9 Flagella and Virulence in Bacteria

In *V. cholerae* motility and chemotactic ability contribute to virulence, as tested in animal models, presumably because they help the bacteria to reach sites for attachment. There has been little evidence that chemotaxis, or motility alone, contributes to virulence in the Enterobacteriaceae. However, recent reports show that for Ity-s mice the virulence, by the oral, i.v., or i.p. route, of an *S. typhimurium* strain was substantially reduced by introduction of a *flaF* allele (by cotransduction with a silent insertion of transposon Tn10, causing absence of flagella (CARSIOTIS et al. 1984), and that *flaF25* or *fla*$^+$ character, determining presence or absence of flagella, affected the ability of the bacteria to survive and multiply in macrophages in vitro (WEINSTEIN et al. 1984). Making the same strain chemotaxis-deficient, or "paralyzed," by transducing in mutant *che* or *mot* alleles, by contrast had no effect on virulence. The way in which the *fla* allele affected fate in macrophages is not known. No inference as to the role of flagella in virulence of *Salmonella* sp. should be drawn until the effect of other *fla* alleles has been tested.

10 Conclusion

Does the genetic determination of properties affecting bacterial virulence or pathogenesis differ in any way from that of other bacterial properties? One striking feature is the polymorphism, in the sense of presence in a genus or other taxon of many functionally equivalent genes or gene clusters determining antigenically (and therefore structurally) different forms of some virulence-related surface component; another is versatility, i.e., the ability of bacteria of a given strain or clone to express any one of two or more alternative forms of such a surface structure. There seems no reason to doubt that the versatility of surface antigen composition of *Borrelia recurrentis* was evolved as a reply to the ability of the host to make antibody against, and thus eliminate, the antigenic form prevalent at any one time in an infection. Host immune responses may also account both for the polymorphism of O (LPS) and H (flagellar) antigens in *Salmonella*, etc., and for the versatility (form variation, phase variation) of these structures in individual *Salmonella* species (though most evidence suggests that host immune response to a given flagellar antigen has little if any effect in curtailing a generalized infection or in preventing a new infection by a strain with the flagellar antigen in question). However, one should bear in mind that a generally similar range of O and H antigens are met in genera, such as *Proteus*, in which host immune responses would not be expected to play any important evolutionary role. In such genera (at least) it may be that polymorphism and perhaps also versatility reflect genetic drift or selective pressures exerted by unknown environmental features.

It is also noteworthy that many virulence-related characters are determined by plasmid or prophage genes, rather than by chromosomal genes. This may reflect the facility with which some newly evolved determinant conferring selective advantage to a pathogenic species can spread by horizontal transmission. In this respect the development of nonchromosomal determinants of pathogenic properties resembles the much more rapid evolutionary development of plasmids (and prophages) controlling resistance to antimicrobial substances, observed during the 50 years or so in which the agents concerned have been in use, in medicine, agriculture, etc.

Acknowledgments: The work of B.A.D.S. was supported by U.S. Public Health Service Research Grants AI 07168 and AI 18872 from the National Institute of Allergy and Infections Diseases.

References

Anderson RP, Roth JR (1977) Tandem genetic duplications in phage and bacteria. Annu Rev Microbiol 31:473–505

Babior BM (1978) Oxygen-dependent microbial killing by phagocytes. New Engl J Med 298:659–668; 721–725

Bacon GA, Burrows TW, Yates M (1950a) The effects of biochemical mutation on the virulence of *Bacterium typhosum*: the induction and isolation of mutants. Br J Exp Pathol 31:703–713

Bacon GA, Burrows TW, Yates M (1950b) The effect of biochemical mutation on the virulence of *Bacterium typhosum*: the virulence of mutants. Br J Exp Pathol 31:714–724

Bacon GA, Burrows TW, Yates M (1951) The effects of biochemical mutation on the virulence of *Bacterium typhosum*: the loss of virulence of certain mutants. Br J Exp Pathol 32:85–96

Benjamin WH Jr, Turnbough CL Jr, Posey BS, Briles DE (1985) The ability of *Salmonella typhimurium* to produce the iron gathering siderophore, enterobactin, is not a virulence factor in mouse typhoid. Infect Immun 50:392–397

Berche PA, Carter PB (1982) Calcium requirement and virulence of *Yersinia enterocolitica*. J Med Microbiol 15:277–284

Bhakdi S, Tranum-Jensen J, Klump O (1980) The terminal membrane C5b-9 complex of human complement. Evidence for the existence of multiple protease-resistant polypeptides that form the trans-membrane complement channel. J Immunol 124:2451–2457

Björnson AB, Björnson HS (1977) Activation of complement by opportunist pathogens and chemotypes of *Salmonella minnesota*. Infect Immun 16:748–753

Blanden RV, MacKaness GB, Collins FM (1966) Mechanisms of acquired resistance in mouse typhoid. J Exp Med 124:585–600

Blumenstock E, Jann K (1981) Natural resistance of mice to *Salmonella typhimurium*: bactericidal activity and chemiluminescence response of murine peritoneal macrophages. J Gen Microbiol 125:173–179

Bölin I, Norlander L, Wolf-Watz H (1982) Temperature-inducible outer membrane protein of *Yersinia pseudotuberculosis* and *Yersinia enterocolitica* is associated with the virulence plasmid. Infect Immun 37:506–512

Brinton CC (1959) Non-flagellar appendages of bacteria. Nature 183:782–786

Brubaker RR (1972) The genus *Yersinia*: biochemistry and genetics of virulence. Curr Top Microbiol Immunol 57:111–158

Carsiotis M, Weinstein DL, Karch H, Holder IA, O'Brien AD (1984) Flagella of *Salmonella typhimurium* are a virulence factor in infected C57BL/6J mice. Infect Immun 46:814–818

Clegg S (1982) Cloning of genes determining the production of mannose-resistant fimbriae in a uropathogenic strain of *Escherichia coli* belonging to serogroup O6. Infect Immun 38:739–744

Coleman W, Leive L (1979) Two mutations which affect the barrier function of the *Escherichia coli* K-12 outer membrane. J Bacteriol 139:899–910

Collins FM (1969) Effect of immune mouse serum on the growth of *Salmonella enteritidis* in non-vaccinated mice challenged by various routes. J Bacteriol 97:667–675

Cooper GN, Fahey KJ (1970) Oral immunization in experimental salmonellosis. III. Behavior of virulent and temperature-sensitive mutant strains in the intestinal tissues of rats. Infect Immun 2:192–200

de Graaf FK, Klaasen-Boor P, van Hees JE (1980) Biosynthesis of the K99 surface antigen is repressed by alanine. Infect Immun 30:125–128

Droge W, Ruschman E, Lüderitz O, Westphal O (1968) Biochemical studies on lipopolysaccharides of *Salmonella* R mutants. 4. Phosphate groups linked to heptose units and their absence in some R lipopolysaccharides. Eur J Biochem 4:134–138

Duguid JP, Gillies RR (1958) Fimbriae and haemagglutinating activity in *Salmonella, Klebsiella, Proteus* and *Chromobacterium*. J Pathol Bacteriol 75:519–520

Duguid JP, Darekar MR, Wheater DWF (1976) Fimbriae and infectivity in *Salmonella typhimurium*. J Med Microbiol 9:459–473

Edwang TG, Befus AD (1984) The role of complement in the induction and regulation of immune responses. Immunology 51:207–224

Eisenstein B (1981) Phase variation of Type 1 fimbriae in *Escherichia coli* is under transcriptional control. Science 214:337–339

Elsbach P, Weiss J (1983) A reevaluation of the roles of the O_2-dependent and O_2 independent microbicidal systems of phagocytes. Rev Infect Dis 5:843–853

Finne J, Leinonen M, Makela PH (1983) Antigenic similarities between brain components and bacteria causing meningitis. Lancet 2:354–357

Gaastra W, de Graaf FK (1982) Host-specific fimbrial adhesins of noninvasive enterotoxigenic *Escherichia coli* strains. Microbiol Rev 46:129–161

Gemski P, Lazere JR, Casey T (1980) Plasmid associated with pathogenicity and calcium dependency of *Yersinia enterocolitica*. Infect Immun 27:682–685

Girardeau JP, Dubourguier HC, Gouet Ph (1982) Effect of glucose and amino acids on expression of K99 antigen in *Escherichia coli*. J Gen Microbiol 128:2243–2249

Göransson M, Uhlin BE (1984) Environmental temperature regulates transcription of a virulence pili operon in *E. coli*. EMBO J 3:2885–2888

Griffin FM Jr, Mullinax PJ (1981) Augmentation of macrophage complement receptor function in vitro. III. C3b receptors that promote phagocytosis migrate within the plane of the macrophage plasma membrane. J Exp Med 154:291–305

Grossman N, Leive L (1984) Complement activation via the alternative pathway by purified *Salmonella* lipopolysaccharide is affected by its structure but not its O-antigen length. J Immunol 132:376–385

Gutteridge WE, Coombes GH (1977) Biochemistry of parasitic protozoa. University Park Press, Baltimore

Hackstadt T, Williams JC (1981) Biochemical stratagem for obligate parasitism of eukaryotic cells by *Coxiella burnetii*. Proc Natl Acad Sci USA 78:3240–3244

Halula M, Stocker BAD (1984) Cloning mannose-resistant hemagglutination gene(s) in *Salmonella typhimurium*. Abstr Annu Meet Am Soc Microbiol p 24

Halula M, Stocker BAD (1985) Mannose-resistant haemagglutination gene(s) of *Salmonella typhimurium*. Abstr Annu Meet Am Soc Microbiol p 34

Hammer CH, Shin ML, Abramovitz AS, Mayer MM (1977) On the mechanism of cell membrane damage by complement: evidence on insertion of polypeptide chains from C8 and C9 into the lipid bilayer of erythrocytes. J Immunol 119:1–8

Herzberg M (1962) Living organisms as immunizing agents against experimental salmonellosis in mice. I. Virulence of auxotrophic mutants. J Infect Dis 111:192–203

Hoiseth SK, Stocker BAD (1981) Aromatic-dependent *Salmonella typhimurium* are non-virulent and effective as live vaccines. Nature 291:238–239

Hull RA, Gill RE, Hsu P, Minshew BH, Falkow S (1981) Construction and expression of recombinant plasmids encoding type 1 or D-mannose-resistant pili from a urinary tract infection *Escherichia coli*. Infect Immun 33:933–938

Ivanovics G, Marjai E, Dobozy A (1968) The growth of purine mutants of *Bacillus anthracis* in the body of the mouse. J Gen Microbiol 53:147–162

Joiner KA, Hammer CH, Brown EJ, Cole RJ, Frank MM (1982a) Studies on the mechanism of bacterial resistance to complement-mediated killing. I. Terminal complement components are deposited and released from *Salmonella minnesota* S 218 without causing bacterial death. J Exp Med 155:797–808

Joiner KA, Hammer CH, Brown EJ, Frank MM (1982b) Studies on the mechanism of bacterial resistance to complement-mediated killing. II. C8 and C9 release C5b67 from the surface of *Salmonella minnesota* S 218 because the terminal complex does not insert into the bacterial outer membrane. J Exp Med 155:809–819

Joiner KA, Schmetz MA, Goldman RC, Leive L, Frank MM (1984) Mechanism of bacterial resistance to complement-mediated killing: inserted C5b-9 correlates with killing for *Escherichia coli* O111B4 varying in O-antigen capsule and O-polysaccharide coverage of lipid A core oligosaccharide. Infect Immun 45:113–117

Jones GW, Rutter JM (1972) Role of the K88 antigen in the pathogenesis of neonatal diarrhea caused by *Escherichia coli* in piglets. Infect Immun 6:918–927

Jones GW, Rabert DK, Svinarich DM, Whitfield HJ (1982) Association of adhesive, invasive, and virulent phenotypes of *Salmonella typhimurium* with autonomous 60-megadalton plasmids. Infect Immun 38:476–486

Källenius G, Mollby R, Svenson SB, Winberg J, Hultberg H (1980) Identification of a carbohydrate receptor recognized by uropathogenic *Escherichia coli*. Infection 8:S 288–S 293

Kauffmann F (1941) A typhoid variant and a new serological variation in the *Salmonella* group. J Bacteriol 41:127–140

Korhonen TK, Valtonen MV, Parkkinen J, Väisänen-Rhen V, Finne J, Ørskov F, Ørskov I, Svenson SB, Mäkelä PH (1985) *Escherichia coli* strains associated with neonatal sepsis and meningitis: serotypes, hemolysin production and receptor recognition. Infect Immun 48:386–491

Labigne-Roussel AF, Lark D, Schoolnik G, Falkow S (1984) Cloning and expression of an afimbrial adhesin (AFA-I) responsible for P blood group-independent, mannose-resistant hemagglutination from a pyelonephritic *Escherichia coli* strain. Infect Immun 46:251–259

Langenberg M-L, Tytgat GNJ, Schipper MEI, Rietra PJGM, Zanen HC (1984) *Campylobacter*-like organisms in the stomach of patients and healthy individuals. Lancet 1:1348

Law SK, Lichtenberg NA, Levine RP (1979) Evidence for an ester linkage between the labile binding site of C3b and receptive surfaces. J Immunol 123:1388–1394

Lederberg J, Iino T (1956) Phase variation in *Salmonella*. Genetics 41:744–757

Leffler H, Svanborg Eden C (1980) Chemical identification of a glycosphingolipid receptor for *Escherichia coli* attaching to human urinary tract epithelial cells and agglutinating human erythrocytes. FEMS Microbiol Lett 8:127–134

Liang-Takasaki C-J, Mäkelä PH, Leive L (1982) Phagocytosis of bacteria by macrophages: changing the carbohydrate of lipopolysaccharide alters interaction with complement and macrophages. J Immunol 128:1229–1235

Liang-Takasaki C-J, Grossman N, Leive L (1983) Salmonellae activate complement differentially via the alternative pathway depending on the structure of their lipopolysaccharide O-antigen. J Immunol 130:1867–1870

Linde K, Keller H, Ezold R, Blatz B, Gericke B, Koch H, Kittlick M, Schmidt S (1974) Live vaccines against infections with Enterobacteriaceae: problems of selection of attenuated mutants and their genetic stability. Acta Microbiol Acad Sci Hung 21:11–27

Lysko PG, Morse SA (1981) *Neisseria gonorrhoeae* cell envelope: permeability to hydrophobic molecules. J Bacteriol 145:946–952

MacKaness GB, Blanden RV, Collins FM (1966) Host-parasite relations in mouse typhoid. J Exp Med 124:573–583

Mäkelä PH, Mäkelä O (1966) *Salmonella* antigen 12_2: genetics of form variation. Ann Med Exp Fenn 44:310–317

Mäkelä PH, Stocker BAD (1984) Genetics of lipopolysaccharide. In: Rietschel ET (ed) Handbook of endotoxin, vol 1: chemistry of endotoxin. Elsevier, Amsterdam, pp 50–137

Mekalanos JJ (1983) Duplication and amplification of toxin genes in *Vibrio cholerae*. Cell 35:253–263

Modrzakowski MC, Spitznagel JK (1979) Bactericidal activity of fractionated granule contents from human polymorphonuclear leukocytes: antagonism of granule cationic proteins by lipopolysaccharide. Infect Immun 25:597–602

Moll A, Manning PA, Timmis KN (1980) Plasmid-determined resistance to serum bactericidal activity: a major outer membrane protein, the *traT* gene product, is responsible for plasmid-specified serum resistance in *Escherichia coli*. Infect Immun 28:359–367

Moon HW, Nagy B, Isaacson RE, Ørskov I (1977) Occurrence of K99 antigen on *Escherichia coli* isolated from pigs and colonization of pig ileum by K99$^+$ enterotoxigenic *Escherichia coli* from calves and pigs. Infect Immun 15:614–620

Morrison DC, Kline LF (1977) Activation of the classical and properdin pathways of complement by bacterial lipopolysaccharides (LPS). J Immunol 118:362–368

Nevola JJ, Stocker BAD, Laux DC, Cohen PS (1985) Colonization of the mouse intestine by an avirulent *Salmonella typhimurium* strain and its lipopolysaccharide-defective mutants. Infect Immun 50:152–159

Normark S (1969) Mutation in *Escherichia coli* K-12 mediating spherelike envelopes and changed tolerance to ultraviolet irradiation and some antibiotics. J Bacteriol 98:1274–1277

Normark S, Lark D, Hull R, Norgren M, Baga M, O'Hanley P, Schoolnik G, Falkow S (1983) Genetics of digalactoside-binding adhesin from uropathogenic *Escherichia coli* strain. Infect Immun 41:942–949

Normark S, Båga M, Göransson M, Lindberg F, Lund B, Norgren M, Uhlin B-E (1985) Genetics of bacterial adhesins. In: Korhonen TK, Dawes EA, Mäkelä PH (eds) Enterobacterial surface antigens: methods for molecular characterization. Elsevier, Amsterdam

Nowicki B, Rhen M, Väisänen-Rhen V, Pere A, Korhonen TK (1984) Immunofluorescence study of fimbrial phase variation in *Escherichia coli* KS71. J Bacteriol 160:691–695

O'Brien AD, Scher I, Formal SB (1979) Effect of silica on the innate resistance of inbred mice to *Salmonella typhimurium* infection. Infect Immun 25:513–520

O'Brien AD, Newland JW, Miller SF, Holmes RK (1984) *Shiga*-like toxin-converting phages from *Escherichia coli* strains that cause hemorrhagic colitis or infantile diarrhea. Science 226:694–696

Old DC (1972) Inhibition of the interaction between fimbrial haemagglutinins and erythrocytes by D-mannose and other carbohydrates. J Gen Microbiol 71:149–157

Orndorff PE, Falkow S (1984) Identification and characterization of a gene product that regulates Type 1 piliation in *Escherichia coli*. J Bacteriol 160:61–66

Ørskov F, Ørskov I, Sutton A, Schneerson R, Lin W, Egan W, Hoff GE, Robbins JB (1979) Form variation in *Escherichia coli* K1: determined by O-acetylation of the capsular polysaccharide. J Exp Med 149:669–685

Ørskov I, Ferenc A, Ørskov F (1980) Tamm-Horsfall protein or uromucoid is the normal urinary slime that traps type 1 fimbriated *Escherichia coli*. Lancet I:887

Pangburn MK, Müller-Eberhard HJ (1980) Relation of a putative thioester bond in C3 to activation of the alternative pathway and the binding of C3b to biological targets of complement. J Exp Med 152:1102

Parkkinen J, Finne J, Achtman M, Väisänen V, Korhonen TK (1983) *Escherichia coli* strains binding neuraminyl 2-3 galactosides. Biochem Biophys Res Commun 111:456–461

Pluschke G, Achtman M (1984) Degree of antibody-independent activation of the classical complement pathway by K1 *Escherichia coli* differs with O antigen type and correlates with virulence of meningitis in newborns. Infect Immun 43:684–692

Pluschke G, Mayden J, Achtman M, Levine RP (1983a) Role of the capsule and the O antigen in resistance of O18:K1 *Escherichia coli* to complement-mediated killing. Infect Immun 42:907–913

Pluschke G, Mercer A, Kusecek B, Pohl A, Achtman M (1983b) Induction of bacteremia in newborn rats by *Escherichia coli* K1 is correlated with only certain O (lipopolysaccharide) antigen types. Infect Immun 39:599–608

Portnoy DA, Blank HF, Kingsbury DT, Falkow S (1983) Genetic analysis of essential plasmid determinants of pathogenicity in *Yersinia pestis*. J Infect Dis 148:297–304

Rhen M, Knowles J, Penttilä ME, Sarvas M, Korhonen TK (1983a) P fimbriae of *Escherichia coli*: molecular cloning of DNA fragments containing the structural genes. FEMS Microbiol Lett 19:119–123

Rhen M, Mäkelä PH, Korhonen TK (1983b) P fimbriae of *Escherichia coli* are subject to phase variation. FEMS Microbiol Lett 19:267–271

Robertsson JA, Lindberg AA, Hoiseth SK, Stocker BAD (1983) *Salmonella typhimurium* infection in calves: evaluation of protection and survival of virulent *S. typhimurium* challenge bacteria after immunization with live or inactivated *S. typhimurium* vaccines. Infect Immun 41:742–750

Rutter JM, Jones GW (1973) Protection against enteric disease caused by *Escherichia coli*: a model for vaccination with a virulence determinant. Nature 242:531–533

Saier MH, Schmidt MR, Leibowitz M (1978) Cyclic AMP-dependent synthesis of fimbriae in *Salmonella typhimurium*: effects of *cya* and *pts* mutations. J Bacteriol 134:356–358

Sansonetti PJ, Kopecko DJ, Formal SB (1982) Involvement of a plasmid in the invasive ability of *Shigella flexneri*. Infect Immun 35:852–860

Sansonetti PJ, Hale TL, Dammin GJ, Kapfer C, Collins HH, Formal SB (1983) Alterations in the pathogenicity of *Escherichia coli* K-12 after transfer of plasmid and chromosomal genes from *Shigella flexneri*. Infect Immun 39:1392–1402

Saxen H (1984) Mechanism of the protective action of anti-*Salmonella* IgM in experimental mouse salmonellosis. J Gen Microbiol 130:2277–2283

Saxen H, Hovi M, Mäkelä PH (1984) Lipopolysaccharide and mouse virulence of *Salmonella*: O antigen is important after intraperitoneal but not intravenous challenge. FEMS Microbiol Lett 24:63–66

Schlecht S, Schmidt G (1969) Möglichkeiten zur Differenzierung von *Salmonella*-R-Formen mittels Antibiotica und antibakterieller Farbstoffe. Zentralbl Bakteriol Parasitenk Infektionskr Hyg Abt 1, Orig. 212:505–511

Schreiber RD, Morrison DC, Podack ER, Müller-Eberhard J (1979) Bactericidal activity of the alternative complement pathway generated from 11 isolated plasma proteins. J Exp Med 149:870–882

Schweizer M, Schwarz H, Sonntag I, Henning U (1976) Mutational change of membrane architecture. Mutants of *Escherichia coli* K12 missing major proteins of the outer cell envelope membrane. Biochim Biophys Acta 448:474–491

Sellwood R, Gibbons RA, Jones GW, Rutter JM (1975) Adhesion of enteropathogenic *Escherichia coli* to pig intestinal brush borders: the existence of two pig phenotypes. J Med Microbiol 8:405–411

Simonet M, Mazigh D, Berche P (1984) Growth of *Yersinia pseudotuberculosis* in mouse spleen despite loss of a virulence plasmid of mol. wt. 47×10^6. J Med Microbiol 18:371–375

Skurnik M, Bölin I, Heikkinen H, Piha S, Wolf-Watz H (1984) Virulence plasmid-associated autoagglutination in *Yersinia* spp. J Bacteriol 158:1033–1036

Smith BP, Reina-Guerra M, Hoiseth SK, Stocker BAD, Habasha F, Johnson E, Meritt F (1983) Safety and efficacy of aromatic-dependent *Salmonella typhimurium* as live vaccine for calves. Am J Vet Res 45:59–66

Smith BP, Reina-Guerra M, Stocker BAD, Hoiseth SK, Johnson E (1984) Aromatic-dependent *Salmonella dublin* as a parenteral modified live vaccine for calves. Am J Vet Res 45:2231–2235

Smith HW, Huggins MB (1980) The association of the O18, K1 and H7 antigens and the ColV plasmid of a strain of *Escherichia coli* with its virulence and immunogenicity. J Gen Microbiol 121:387–400

Smith HW, Linggood MA (1971) Observations on the pathogenic properties of the K88 *hly* and *ent* plasmids of *Escherichia coli* with particular reference to porcine diarrhea. J Med Microbiol 4:467–485

Smith HW, Tucker JF (1976) The virulence of trimethoprim-resistant thymine-requiring strains of *Salmonella*. J Hyg 76:97–108

Stevens P, Young LS, Adamu S (1983) Opsonization of various capsular (K) *E. coli* by the alternative complement pathway. Immunology 50:497–502

Stocker BAD, Hoiseth SK, Smith BP (1983) Aromatic-dependent *Salmonella* sp. as live vaccine, in mice and calves. Dev Biol Stand 53:47–62

Sukupolvi S, Vaara M, Helander IM, Viljanen P, Mäkelä PH (1984) New *Salmonella typhimurium* mutants with altered outer membrane permeability. J Bacteriol 159:704–712

Takeuchi A (1967) Electron microscope studies of experimental *Salmonella* infection. I. Penetration into the intestinal epithelium by *Salmonella typhimurium*. Am J Pathol 50:109–136

Tenner AJ, Ziccardi RJ, Cooper NR (1984) Antibody-independent C1 activation by *E. coli*. J Immunol 133:886–891

Tidmarsh GF, Rosenberg LT (1981) Aquisition of iron from transferrin by *Salmonella paratyphi* B. Current Microbiol 6:217–220

Vaara M, Nikaido H (1984) Molecular organization of bacterial outer membrane. In: Rietschel ET (ed) Handbook of endotoxin, vol 1: chemistry of endotoxin. Elsevier, Amsterdam, pp 1–45

Vaara M, Vaara T (1983) Sensitization of Gram-negative bacteria to antibiotics and complement by a nontoxic oligopeptide. Nature 133:526–527

Vaara M, Viljanen P, Vaara T, Mäkelä PH (1984) An outer membrane-disorganizing peptide PMBN sensitizes *E. coli* strains to serum bactericidal action. J Immunol 132:2582–2589

Väisänen V, Elo J, Tallgren LG, Siitonen A, Mäkelä PH, Svanborg Eden C, Källenius G, Svenson SB, Hultberg H, Korhonen TK (1981) Mannose-resistant hemagglutination and P antigen recognition are characteristic of *Escherichia coli* causing primary pyelonephritis. Lancet II:1366–1369

Valtonen MV (1977) Role of phagocytosis in mouse virulence of *Salmonella typhimurium* recombinants with O-antigen 6, 7 or 4, 12. Infect Immun 18:574–578

Valtonen VV (1970) Mouse virulence of *Salmonella* strains: the effect of different smooth-type O side-chains. J Gen Microbiol 64:255–268

Weinstein DL, Carsiotis M, Lissner CR, O'Brien AD (1984) Flagella help *Salmonella typhimurium* survive within murine macrophages. Infect Immun 46:819–825

Weiss J, Victor M, Elsbach P (1983) Role of charge and hydrophobic interaction in the action of the bactericidal/permeability-increasing protein of neutrophils on gram-negative bacteria. J Clin Invest 71:540–549

Williams PH, Warner PJ (1980) ColV plasmid-mediated, colicin V-independent iron uptake system of invasive strains of *Escherichia coli*. Infect Immun 29:411–416

Wright SD, Levine RP (1981) How complement kills *E. coli*. I. Location of the lethal lesion. J Immunol 127:1146–1151

Yancey RJ, Breeding SAL, Lankford CE (1979) Enterochelin (enterobactin): virulence factor for *Salmonella typhimurium*. Infect Immun 24:174–180

Zaleska M, Lounatmaa K, Nurminen M, Wahlström E, Mäkelä PH (1985) A novel virulence-associated cell surface structure composed of 47 Kdal protein subunits in *Yersinia enterocolitica*. EMBO J 4:1013–1018

Zieg J, Silverman M, Hilmen M, Simon M (1977) Recombinational switch for gene expression. Science 196:170–172

Zinc DL, Feeley JC, Wells JG, Vanderzant C, Vickery JC, Roof WD, O'Donoran GA (1980) Plasmid-mediated tissue invasiveness in *Yersinia enterocolitica*. Nature 283:224–226

---

* The above refers to a preliminary account of the investigation; it is now reported (BENJAMIN et al. 1985) that introduction of the *enb-7* mutation did not detectably reduce virulence, as judged by mortality from small inocula, even when the challenge was i.p., instead of i.v. No explanation for the discrepancy between this result and that of YANCEY et al. (1979) is apparent, other than differences in the strains of mice and of bacteria used in the two investigations.

Subject Index